4th International Conference on Smart Material Research (ICSMR 2018)

AA006908

IOP Conference Series: Materials Science and Engineering
Volume 522

Sydney, Australia
16 – 18 November 2018

ISBN: 978-1-5108-8816-6
ISSN: 1757-8981

Printed from e-media with permission by:

Curran Associates, Inc.
57 Morehouse Lane
Red Hook, NY 12571

Some format issues inherent in the e-media version may also appear in this print version.

This work is licensed under a Creative Commons Attribution 3.0 International Licence. Licence details: http://creativecommons.org/licenses/by/3.0/.

No changes have been made to the content of these proceedings. There may be changes to pagination and minor adjustments for aesthetics.

Printed by Curran Associates, Inc. (2019)

For permission requests, please contact the Institute of Physics
at the address below.

Institute of Physics
Dirac House, Temple Back
Bristol BS1 6BE UK

Phone: 44 1 17 929 7481
Fax: 44 1 17 920 0979

techtracking@iop.org

Additional copies of this publication are available from:

Curran Associates, Inc.
57 Morehouse Lane
Red Hook, NY 12571 USA
Phone: 845-758-0400
Fax: 845-758-2633
Email: curran@proceedings.com
Web: www.proceedings.com

Table of contents

Volume 522

2018 4th International Conference on Smart Material Research

16–18 November 2018, Sydney, Australia

Accepted papers received: 22 March 2019

Preface

2018 4th International Conference on Smart Material Research

Committee List

Peer review statement

Papers

Composites and Film Technology

Sound absorption performance of underwater anechoic coating in plane wave normal incidence condition

Z Zhang, L Li, Y Huang and Q Huang.....1

Preparation and characterization of conductive Polyaniline/Sago/Graphene nanocomposites via ultrasonic irradiation

M E Ali Mohsin, A Arsad, A Hassan, N K Shrivastava and M A Ahmad Zaini.....7

Interface Damage of Protective Layer in TEM Lamella Preparation for Highly Doped Ge Substrate

Nur Nadhirah Mohamad Rashid, Norhamizah Hazirah Ahmad Junaidi and Siti Rahmah Aid.....12

Influence of anneal temperature in air on surface morphology and photoluminescence of ZnO thin films

Sujun Guan, Liang Hao, Mariko Murayama, Xiaohua Xie, Shuji Komuro and Xinwei Zhao.....17

Numerical modelling of pseudoelastic behaviour of geometrically graded shape memory alloys

Bashir S Shariat, Sam Bakhtiari and Yinong Liu.....25

Characterization of the magnetorheological fluid MRF-140 CG for applications in robotic prosthesis

Oscar Arteaga, Marcelo Ortiz, Eduardo Cárdenas, Segundo M. Espin-Lagos, Miguel A. Carvajal, Héctor C. Terán and Mario A. Lara.....30

Optimization of Bilayer Actuator Based on Carbon Black/Polymer Composites

H Shivashankar, R Sangamesh and S M Kulkarni.....35

Analytical and numerical modeling of a clamped-clamped micro beam under electric actuation

Aalim M Mustafa and Ahmed K Mahgoub.....42

Building Materials

Cement Substitutions in UHPC and their Influence on Principal Mechanical-Physical Properties

K Šeps, I Broukalová and R Chylík.....49

Analysis of the influence of supplementary cementitious materials used in UHPC on modulus of elasticity

J Fládr, P Bílý and T Trtík.....53

Effect of GGBS and Nano Silica on the Mechanical Properties of Ternary Concrete

Manank Shah, Niragi Dave, Kaushal Vora and Dhruv Shah.....58

Improving the Resistance of Cement-Based Composites to Sulphur Dioxide

N Žižková, L Nevřivová, M Lédl and Š Keprdová.....63

The Influence of Crystallization Additives on the Porosity of Polypropylene Fibre-reinforced Cement Mortar

L Nevřivová and N Žižková.....69

Numerical Analysis of Solid Fuel Regression Rate Distribution Characteristics in Solid-fuel Ramjet with Different Fuel Types

Omer Musa, Chen Xiong, Liao Wenhe and Li Yingkun.....75

IOP Conf. Series: Materials Science and Engineering **522** (2019) 011001 doi:10.1088/1757-899X/522/1/011001

PREFACE

It is with great pleasure that I introduce the proceedings of the 2018 4th International Conference on Smart Material Research (ICSMR 2018) which was held recently in Sydney, Australia, 16-18, November 2018. ICSMR 2018 was dedicated to technical issues related to smart material research.

The objectives of the conference were not only to provide opportunities to the international scientists to present state of the art research in smart materials but also to provide a forum to discuss ideas for future directions in the field and explore possibilities for collaborative research projects.

The conference program included keynote, oral, and poster presentations from scholars working in the areas of materials science and engineering from all over the world. It covered recent trends and progress made in the field of materials engineering and manufacturing. Professors from Australia and USA were invited to deliver three keynote lectures regarding the latest information in their respective areas of expertise.

The proceedings present a selection of high quality papers submitted to the conference by researchers from universities, research institutes, and industry. All papers were subjected to peer-review by conference committee members and international reviewers. The papers were selected based on their quality and their relevance to the conference. The proceedings present recent advances in the fields of materials science and engineering and manufacturing as well as in other related areas including materials properties, measuring methods, analytical and numerical modeling and simulations among others.

I would like to express special gratitude to members of the conference committee and organizers of the conference. I would also like to thank the reviewers for their valuable time and advice which helped in improving the quality of the papers selected for presentation at the conference and for publication in the proceedings. Finally I want to thank the authors, the members of the organizing committee, the reviewers, the chairpersons, sponsors, and all other conference participants for their support of ICSMR 2018.

Prof. Ramesh K. Agarwal
Washington University in St. Louis, USA
December 21, 2018

Content from this work may be used under the terms of the Creative Commons Attribution 3.0 licence. Any further distribution of this work must maintain attribution to the author(s) and the title of the work, journal citation and DOI.
Published under licence by IOP Publishing Ltd

IOP Conf. Series: Materials Science and Engineering **522** (2019) 011002 doi:10.1088/1757-899X/522/1/011002

Conference Committee Chairs

Darren Martin, University of Queensland, Australia
Ramesh K. Agarwal, Washington University in St. Louis, USA

Program Committee Chairs

Rongkun Zheng, The University of Sydney, Australia
Alexander Pesin, Nosov Magnitogorsk State Technical University, Russia

International Technical Committees

Mike Bambach, The University of Sydney, Australia
Ing Kong, University of Nottingham Malaysia, Malaysia
Kaveh Shahbaz, University of Auckland, New Zealand
Patricia Sanmartín Sánchez, Universidade de Santiago de Compostela, Spain
Chi-wai Kan,The Hong Kong Polytechnic University, Hong Kong
Carlos Rolando Rios-Soberanis, Centro de Investigación Científica de Yucatán, México
Siti Rahmah Aid, Universiti Teknologi Malaysia, Malaysia
AGUS ARSAD, Universiti Teknologi Malaysia, Malaysia
Leslie Joy Diaz, University of the Philippines, Philippines
Mohammadreza Vafaei, Universiti Teknologi Malaysia, Malaysia
Trong-Phuoc Huynh, Can Tho University, Viet Nam
Oguzhan Gunduz, Marmara University, Turkey
G. RAJARAJAN, PGP Educational Institutions, India
Riffat Asim Pasha, Univeristy of Engineering and Technology, Parkistan
Hsaine Zgou, Ibn Zohr University,Morocco
Barnik Saha Roy, National Institute of Technology Agartala, India
Nilofar Asim, National University of Malaysia, Malaysia
Omar Ben Mya, University of ElOued, Algeria

Content from this work may be used under the terms of the Creative Commons Attribution 3.0 licence. Any further distribution of this work must maintain attribution to the author(s) and the title of the work, journal citation and DOI.
Published under licence by IOP Publishing Ltd

IOP Conf. Series: Materials Science and Engineering **522** (2019) 011003 doi:10.1088/1757-899X/522/1/011003

Peer review statement

All papers published in this volume of *IOP Conference Series: Materials Science and Engineering* have been peer reviewed through processes administered by the proceedings Editors. Reviews were conducted by expert referees to the professional and scientific standards expected of a proceedings journal published by IOP Publishing.

Content from this work may be used under the terms of the Creative Commons Attribution 3.0 licence. Any further distribution of this work must maintain attribution to the author(s) and the title of the work, journal citation and DOI.
Published under licence by IOP Publishing Ltd

IOP Conf. Series: Materials Science and Engineering **522** (2019) 012001 doi:10.1088/1757-899X/522/1/012001

Sound absorption performance of underwater anechoic coating in plane wave normal incidence condition

Z Zhang, L Li, Y Huang and Q Huang*

State Key Laboratory of Digital Manufacturing Equipment and Technology, School of Mechanical Science and Engineering, Huazhong University of Science and Technology, Wuhan 430074, PR China

*Corresponding author - qbhuang@hust.edu.cn

Abstract. On account of the theory of non-uniform waveguide and the wave propagation theory in layered media, a complete theoretical model is established for predicting the sound absorption performance of multi-layer underwater composite anechoic coating. Furthermore, the correctness of the theoretical calculation is verified by the finite-element simulation results. Then, the velocity nephograms corresponding the three absorption peak frequencies of the overburden are analyzed to reveal the sound-absorbing mechanism.

1. Introduction

With the increasing demand for acoustic performance of modern underwater equipment, some measures related to vibration damping and noise reduction continue to emerge. Among them, the installation of the underwater anechoic coating on the outer surface of the equipment is an important move, which has the effects of reducing the acoustic target strength of structure itself, suppressing the structural sound radiation, and improving the sound insulation of internal noise. Therefore, it is necessary to study the sound absorption properties of the underwater anechoic overburden.

The research methods for the acoustic properties of sound-absorbing coatings vary depending on their structural forms. For example, Stepanishen and Strozeski [1] analyzed the reflection-transmission performance of a layered viscoelastic lining by transfer matrix method (TMM). Gao et al. [2] proposed a Generalized Multiscale Finite-Element Method (GMsFEM) to study the wave propagation characteristics in non-homogeneous coatings. Haberman et al. [3] and Bonfoh et al. [4] studied the material and acoustic properties of periodically scattering composite sound-absorbing structures containing spherical inclusions and ellipsoidal coated particles, respectively, through self-consistent theory and generalized self-consistent theory. Tao [5] established a complete two-dimensional theoretical model for predicting the acoustic performances of an overburden embracing periodic cylindrical or conical cavities. In addition, for anechoic coatings with complex cavity structures such as spherical, ellipsoidal, super-ellipsoid and non-spherical cavums, Ivansson [6] investigated the sound-deadening effect by finite element method (FEM). Furthermore, Zhong et al. [7] also adopted the FEM to explore the low-frequency sound-absorbing property of a viscoelastic overburden embodying periodic horizontal cavities with infinitely long eccentric cylinders.

Therefore, this paper proposes a multi-layer underwater composite anechoic coating, which is formed by embedding two heterogeneous linings containing periodic cavity structures in two homogeneous plates. Meanwhile, we introduce the theory of non-uniform waveguide and combine the wave propagation theory in layered media to investigate the sound absorption performance of the present coating.

Content from this work may be used under the terms of the Creative Commons Attribution 3.0 licence. Any further distribution of this work must maintain attribution to the author(s) and the title of the work, journal citation and DOI.
Published under licence by IOP Publishing Ltd

2. Theoretical modeling

2.1. Structural elaboration

The underwater anechoic coating shown in Fig. 1 consists of four rubber layers, labeled from top to bottom as layers m with thicknesses h_m (m = 1,2,3,4), respectively, wherein layers 1 and 4 are homogenous linings, and layers 2 and 3 are heterogeneous overburdens containing periodic cavity structures. The coating divides the service water environment into two upper and lower semi-infinite spaces, namely the water and backing conditions in Fig. 1(a), marked as layers 0 and 5, respectively. Moreover, the two diameters, small upper diameter, and large lower diameter, of the truncated cone cavity in layer 2 are respectively r_1 and r_2, while the diameter of the cylindrical cavity in layer 3 is r_2, as illustrated in Fig. 1(b). In addition, the basic structure of the blue dotted square in Fig. 1(b) is a periodic unit having the length-width dimension of l_b. Furthermore, a global Cartesian coordinate system is set up on the upper surface of the coating, where the x- and y-axes are along the array directions of periodic cavities, and the z-axis is perpendicular to the xy-plane downward. When a plane wave having a sound pressure amplitude of p_i impinges on the upper surface of the overburden normally, the reflected and transmitted waves with amplitudes p_r and p_t are derived respectively.

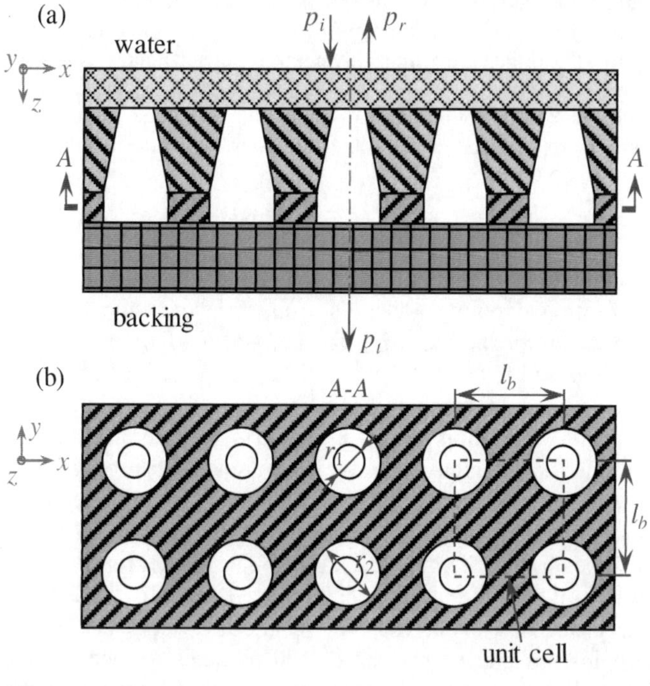

Figure 1. Schematic diagram of the present coating: (a) assembly drawing and (b) sectional view.

2.2. Matrix synthesis and absorption calculation

For a single homogeneous lining similar to layer 1 or 4, the relationship between sound pressure and particle velocity on the front and rear interfaces is [8]:

$$\begin{bmatrix} p_{1,m} \\ v_{1,m} \end{bmatrix} = \begin{bmatrix} \cos(k_m h_m) & \mathrm{i}\rho_m c_m \sin(k_m h_m) \\ \dfrac{\mathrm{i}\sin(k_m h_m)}{\rho_m c_m} & \cos(k_m h_m) \end{bmatrix} \begin{bmatrix} p_{2,m} \\ v_{2,m} \end{bmatrix} = A_m^{\mathrm{pre}} \begin{bmatrix} p_{2,m} \\ v_{2,m} \end{bmatrix}, \quad (1)$$

where subscript m represents the m-th layer in the entire overburden, while subscript 1 and 2 indicate the front and rear ends of the coating; superscript pre denotes the pressure state; ρ, c, h, k, A are respectively the density, complex sound velocity, thickness, complex wavenumber, and transfer matrix of the homogeneous layer; i ($\mathrm{i}^2 = -1$) is the imaginary unit.

Thus, the transfer matrix of a homogeneous layer in the force state is

$$A_m = S_e \otimes A_m^{\text{pre}}, \tag{2}$$

with

$$S_e = \begin{bmatrix} 1 & s_b \\ 1/s_b & 1 \end{bmatrix}, \tag{3}$$

where symbol $\otimes$ indicates the Kronecker product between matrices, $s_b = l_b^2$ is the sectional area of one periodic element.

In layer 2 with periodic truncated cone cavities, the propagation of sound waves in the coating is similar to that in a highly viscous liquid with variable cross-section waveguides, so the wave equation is [9]

$$\frac{\partial^2 \xi}{\partial z^2} + \frac{1}{s}\frac{\partial s}{\partial z}\frac{\partial \xi}{\partial z} + k^2 \xi = 0, \tag{4}$$

where ξ and s are the particle displacement and cross-sectional area of the coating, respectively, and both are all function of z. Therefore, the expression is a nonlinear equation, which can be transformed into a linear equation to obtain the analytic solution only when the medium cross-sectional area $s(z)$ satisfies the following formula:

$$\left(\sqrt{s}\right)'' \Big/ \sqrt{s} = \mu^2 \quad (\mu = \text{constant}). \tag{5}$$

Since the total pressure and particle velocity on the front and rear interfaces in the heterogeneous layer 2 are continuous, the cone-shaped waveguide $\sqrt{s(z)} = az + b$ is used to approximate the truncated cone cavity, and the transfer relations of F_1, F_2, v_1, and v_2 can be expressed as

$$\begin{bmatrix} F_1 \\ v_1 \end{bmatrix} = \begin{bmatrix} b_{2,11} & b_{2,12} \\ b_{2,21} & b_{2,22} \end{bmatrix} = B_2 \begin{bmatrix} F_2 \\ v_2 \end{bmatrix}, \tag{6}$$

with

$$\begin{aligned}
b_{2,11} &= \sqrt{\frac{s_1}{s_2}}\left[\cos(K_2h_2) + \frac{1}{K_2h_2}\left(\sqrt{\frac{s_2}{s_1}} - 1\right)\sin(K_2h_2)\right] \\
b_{2,12} &= \frac{\mathrm{i}\rho_2 c_2 \sqrt{s_1 s_2}}{k_2}\left\{\frac{\cos(K_2h_2)}{h_2}\left(2 - \sqrt{\frac{s_1}{s_2}} - \sqrt{\frac{s_2}{s_1}}\right) + K_2 \sin(K_2h_2)\left[1 - \left(\frac{1}{K_2h_2}\right)^2\left(1 - \sqrt{\frac{s_1}{s_2}}\right)\right]\right\} \\
b_{2,21} &= \frac{\mathrm{i}k_2}{K_2\rho_2 c_2\sqrt{s_1 s_2}}\sin(K_2h_2) \\
b_{2,22} &= \sqrt{\frac{s_2}{s_1}}\left[\cos(K_2h_2) - \frac{1}{K_2h_2}\left(1 - \sqrt{\frac{s_1}{s_2}}\right)\sin(K_2h_2)\right]
\end{aligned}, \tag{7}$$

where $K_2 = \sqrt{k_2^2 - \mu^2}$; $s_1 = l_b^2 - \pi r_1^2$ and $s_2 = l_b^2 - \pi r_2^2$ are the upper and lower interface cross-sectional areas of layer, respectively.

However, for the inhomogeneous layer 3 containing periodical cavities, which is a special case of layer 2, with $s_1 = s_2 = l_b^2 - \pi r_2^2$ and $K_3 = k_3$, so the transfer matrix of sound pressure and particle velocity on both end faces of layer 3 is

$$B_3 = \begin{bmatrix} \cos(k_3h_3) & \mathrm{i}\rho_3 c_3 s_2 \sin(k_3h_3) \\ \dfrac{\mathrm{i}\sin(k_3h_3)}{\rho_3 c_3 s_2} & \cos(k_3h_3) \end{bmatrix}, \tag{8}$$

Moreover, the global transfer matrix of the present coating in the force state can be assembled as follows:

IOP Conf. Series: Materials Science and Engineering **522** (2019) 012001 doi:10.1088/1757-899X/522/1/012001

$$T = A_1(h_1)\times B_2(h_2)\times B_3(h_3)\times A_4(h_4). \tag{9}$$

Then, the reflection and absorption coefficients at the incident interface of the coating are:

$$r = \frac{Z_{in} - Z_w}{Z_{in} + Z_w}, \tag{10}$$

$$\alpha = 1 - r \cdot r^*, \tag{11}$$

with

$$Z_{in} = \frac{T_{11}F_5 + T_{12}v_5}{s_b\left(T_{21}F_5 + T_{22}v_5\right)}, \tag{12}$$

where superscript * denotes the conjugate of the reflection coefficient, $Z_w = \rho_w c_w$ is the characteristic impedance of the aqueous medium in the 0-th layer.

3. Model validations and discussions

3.1. Simulation verification

The value of the present theoretical model is determined by both the calculation accuracy and computation rate, in which the model validity is an indispensable basis. For the underwater anechoic coating as shown in Fig. 1, the materials of layers 1 and 4 are selected as high-elastic rubber (HER), with densities $\rho_1 = \rho_4 = 1300$ kg/m^3, Young's moduli $E_1 = E_4 = 4.96\times10^7(1+i)$ Pa, Poisson ratios $\upsilon_1 = \upsilon_2 = 0.47$, and thicknesses $h_1 = 0.5h_4 = 15$ mm. Moreover, the material properties of ordinary conventional rubber (OCR) in layers 2 and 3 are $\rho_2 = \rho_3 = 1600$ kg/m^3, $E_2 = E_3 = 2.98\times10^8(1+0.4i)$ Pa, $\upsilon_1 = \upsilon_2 = 0.37$, $h_2 = 3h_3 = 30$ mm, and $r_1 = 0.5r_2 = 6$ mm. The density and sound velocity of water in the incident medium are $\rho_w = 1000$ kg/m^3 and $c_w = 1500$ m/s, respectively, while the backing condition is rigid.

Then, a fully coupled finite-element model of the present anechoic coating shown in Fig. 2 is established by COMSOL software in accordance with the following four steps: first, using the perfectly matched layer (PML) to simulate the semi-infinite water space. Second, employing the Floquet periodicity boundaries on the *x*- and *y*-direction planes in one cell to elaborate an infinite lining. Again, approximating the rigid backing by the fixed constraint boundary with vibration velocity of zero. Finally, treating the fluid-solid interfaces between fluids such as water or air and rubbers by the acoustic-structure boundaries.

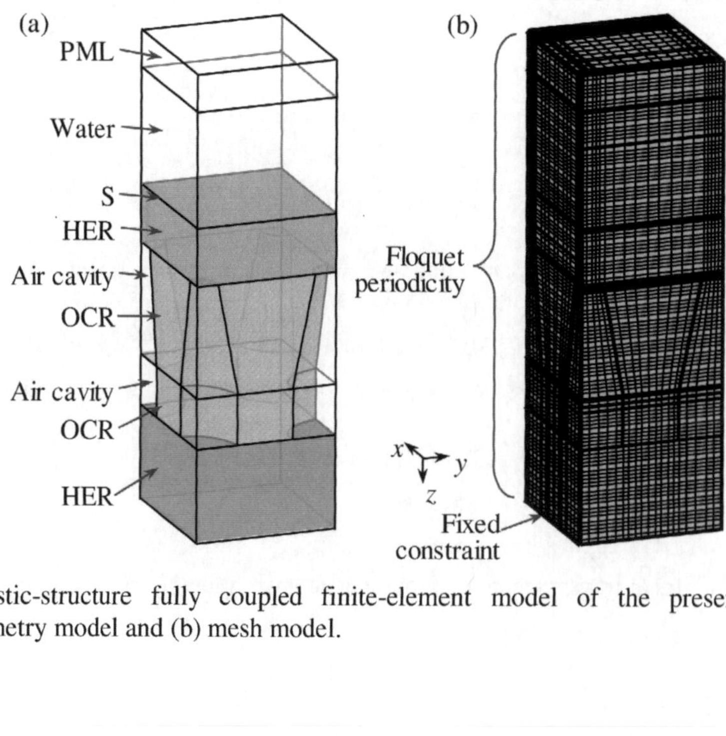

Figure 2. Acoustic-structure fully coupled finite-element model of the present coating: (a) multiphysics geometry model and (b) mesh model.

Therefore, based on the two methods of theoretical calculation and simulation analysis, the sound absorption coefficients of the underwater anechoic overburden are compared as illustrated in Fig. 3. Obviously, the trends of the two curves are consistent, and they all fluctuate with increasing frequency and eventually become stable. In addition, the two highly coincident curves rise rapidly to 0.6, which just shows that the present coating's design is reasonable and the developed theoretical model has higher accuracy.

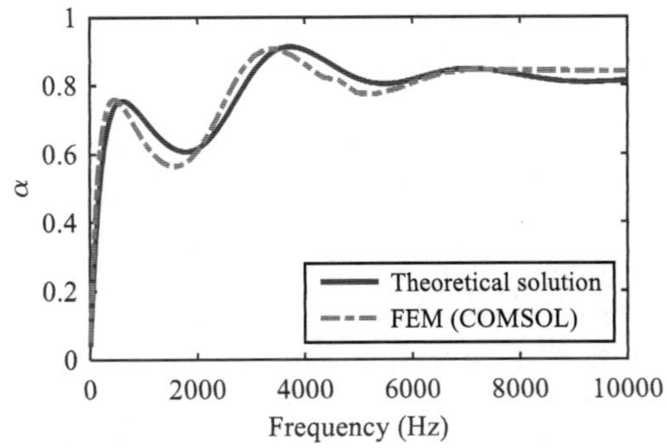

Figure 3. Comparison between the theoretical analysis and simulation result of the present coating.

3.2. Sound-absorbing mechanism exploration

It can be seen from Fig. 3 that the two curves of theoretical calculation and simulation result have three peak frequencies in the research frequency range of 10Hz ~ 10 kHz. In order to explore the energy dissipation mechanisms for the absorbing peaks in the sound absorption curve of the present overburden, the vibration velocity of the coating in the whole research frequency range is simulated and analyzed. The velocity nephograms corresponding to the three peak frequencies are displayed in Fig. 4.

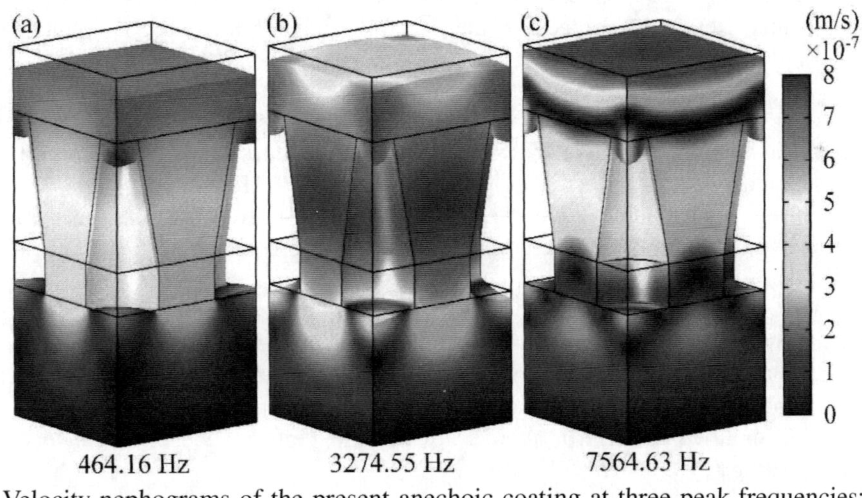

Figure 4. Velocity nephograms of the present anechoic coating at three peak frequencies: (a) 464.16 Hz, (b) 3274.55 Hz and (c) 7564.63 Hz.

From this diagram, we can know that the mechanisms for generating the respective sound absorption peaks are different. For the first peak frequency in the low-frequency band, the sound-absorbing peak at 464.16 Hz is mainly attributable to the axial piston-like resonance energy dissipation generated by the layer 1 coupling the truncated cone cavities in layer 2, as shown in Fig. 4(a), the velocity magnitudes of layers 1 and 2 are dominant. As can be seen from Fig. 4(b), the vibration velocity magnitudes of the heterogeneous layers 2 and 3 are significantly concentrated and superior to other layers. Therefore, the two sound-absorbing modes, the radial drum-like resonance energy consumption of the periodic cavity structures in layers 2 and 3, and the waveform conversion in layer 2, synergistically cause the second sound absorption crests at the frequency 3274.55 Hz in the mid-frequency band. However, for the velocity contour at 7564.63 Hz in the high-frequency range depicted in Fig. 4(c), a number of distinct velocity stratifications are formed within the anechoic coating, while the interface between layer 1 and the aqueous medium is the most intense. This is because the viscous energy dissipation in the anechoic coating plays a major role, so that the sound absorption coefficient of the entire structure is stable in the high-frequency region.

4. Conclusions

This paper concerns with the sound absorption performance of an underwater composite anechoic coating. It is composed of four stratified rubber structures, in which the upper and lower layers are homogeneous linings, while the center two layers are heterogeneous overlays, each containing periodically truncated cone cavums and cylindrical cavities. On account of the theory of non-uniform waveguide and the wave propagation theory in layered media, a complete theoretical model is established for analyzing the acoustic properties of the present coating.

Furthermore, the homogeneous layers 1 and 4 are selected to be high-elastic rubber, and the ordinary conventional rubber is used in place of the non-homogeneous layers 2 and 3, thereby completing the following two aspects of research work. One is to verify the correctness of the developed theoretical model by finite element simulation method. The other is to study and analyze the velocity nephograms corresponding to the three peak frequencies to explore the absorbing mechanisms of the sound absorption crests. Thus, the following two main conclusions are drawn: first, the present theoretical model has higher accuracy. Second, the sound-absorbing mechanisms for the three absorption peaks of the underwater anechoic coating are different, respectively, by the axial piston-like resonance energy dissipation, radial drum-like resonance energy consumption coupled waveform conversion, viscous energy dissipation.

Acknowledgments

This work was supported by the National Natural Science Foundation of China (Grant no. 51575201) and the Fundamental Research Funds for the Central Universities, HUST, China (Grant no. 2018JYCXJJ039).

References

[1] Stepanishen P R and Strozeski B 1982 Reflection and transmission of acoustic wideband plane waves by layered viscoelastic media *J. Acoust. Soc. Am.* **71** 9-21

[2] Gao K, Fu S, Jr R L G, Chung E T and Efendiev Y 2015 Generalized multiscale finite-element method (GMsFEM) for elastic wave propagation in heterogeneous, anisotropic media *J. Comput. Phys.* **295** 161-188

[3] Haberman M R, Berthelot Y H and Jarzynski J 2002 Micromechanical modeling of viscoelastic voided composites in the low-frequency approximation *J. Acoust. Soc. Am.* **112** 1937-1943

[4] Bonfoh N, Hounkpati V and Sabar H 2012 New micromechanical approach of the coated inclusion problem: Exact solution and applications *Comp. Mater. Sci.* **62** 175-183

[5] Tao M 2014 Simplified model for predicting acoustic performance of an underwater sound absorption coating *J. Vib. Control* **20** 339-354

[6] Ivansson S M 2008 Numerical design of Alberich anechoic coatings with superellipsoidal cavities of mixed sizes *J. Acoust. Soc. Am.* **124** 1974-1984

[7] Zhong J, Wen J H, Zhao H G, Yin J F and Yang H B 2015 Effects of core position of locally resonant scatterers on low-frequency acoustic absorption in viscoelastic panel *Chin. Phys. B* **24** 084301

[8] Dupont T, Leclaire P, Panneton R and Umnova O 2018 A microstructure material design for low frequency sound absorption *Appl. Acoust.* **136** 86-93

[9] Du G, Zhu Z and Gong X 2012 *Fundamentals of Acoustics* 3rd ed. (Nanjing, Nanjing University Press) section 5 pp 170-175 [In Chinese]

IOP Conf. Series: Materials Science and Engineering **522** (2019) 012002 doi:10.1088/1757-899X/522/1/012002

Preparation and characterization of conductive Polyaniline/Sago/Graphene nanocomposites via ultrasonic irradiation

M E Ali Mohsin[1], A Arsad[1,2], A Hassan[1], N K Shrivastava[1] and M A Ahmad Zaini[3]

[1] Enhanced Polymer Research Group, School of Chemical and Energy Engineering, Universiti Teknologi Malaysia, 81310, Skudai, Johor, Malaysia
[2] UTM-MPRC Institute for Oil & Gas (IFOG), Universiti Teknologi Malaysia, 81310, Skudai, Johor, Malaysia
[3] Chemical Engineering Department, School of Chemical and Energy Engineering, Universiti Teknologi Malaysia, 81310, Skudai, Johor, Malaysia

agus@cheme.utm.my

Abstract. This work, which is a part of our ongoing studies on developing Polyaniline/Sago/graphene (P/S/G) nanocomposites. The nanocomposites were synthesized via ultrasonic irradiation (U.I) method in presence of hydrochloric acid and Ammonium persulfate. The quantity of PANI and Sago was kept constant and the effects of graphene nanoplatelet (GnP) dosage on their electrical conductivity and morphology was evaluated. Electrical conductivity measurements using a standard four-point probe technique and morphological analysis was carried using Transmission electron microscopy (TEM). The electrical conductivity of nanocomposites was found to be drastically increased as compared to that of PANI/Sago blend at room temperature. Further the conductivity of nanocomposites was also found to be increased with the increase in weight % of GnP. The results also showed that incorporation of less than 1 wt.% of GnP into polymeric matrix had improved electrical conductivity of the nanocomposites from 2.98×10^{-4} to 1.34×10^{-1} S cm^{-1}. The morphology of the generated P/S/G nanocomposites shows uniform dispersion of GnP.

1. Introduction

Conducting polymers, extensively conjugated molecules, have π electron delocalization along their backbone which gives them unique optical and electrical properties. This characteristic causes them to act as a semiconductor or a conductor [1]. Polyaniline (PANI) is one of the most attractive conducting polymers among the known conducting polymers due to some of its characteristics including good environmental, thermal, and chemical stability, tunable conductivity switching between insulating and semiconducting materials, low operational voltage, facile synthesis, potentiality for practical applications and low cost [2]. In recent years, considerable interest has been shown in PANI as a conducting organic polymer because of its potential applications in new technologies, such as chemical sensors, electrochromic displays, electronic devices, and secondary batteries [3-8].
Usually, two methods including chemical oxidation and electrochemical synthesis are employed for preparing PANI. Although prepared PANI via electrochemical synthesis generally has higher conductivity, this method is restricted by the size, shape and nature of the involved electrode. Therefore, this method cannot be used to prepare a large amount of PANI or to produce fine PANI

Content from this work may be used under the terms of the Creative Commons Attribution 3.0 licence. Any further distribution of this work must maintain attribution to the author(s) and the title of the work, journal citation and DOI.
Published under licence by IOP Publishing Ltd

powders; thus PANI used for commercial application is normally synthesized via the chemical oxidation method [9]. However, there are certain limitation of PANI, PANI is non-biodegradable, non-flexible and non-processable, only very low amount of PANI can be and should be used for producing conductive nanocomposites [10]. In addition, the particle size of PANI should be controlled within a proper range, i.e. less than 100 nm, in order that PANI particles can be easily excreted through circulatory systems after complete degradation of polymeric matrix.

The processability of PANI may be achieved by means of several methods that were summarized by Bhadra et al. [11]. Those methods include: chemical modification of aromatic rings or alkylation of nitrogen atoms, modification of doping process using specific Broensted or Lewis acids, or synthesis of the colloidal form of PANI. The second group of processability improvements belongs to the group of PANI composites. According to that, PANI can be applied in a suitable insulating matrix as conductive filler. These types of composites can be easily processed and they show quite good mechanical properties. Today, several preparation methods of PANI composites are known. One of them is in situ polymerization. Three procedures of such method are known: simultaneous synthesis of both PANI and a matrix partner in the same medium; polymerization of aniline in a suitable counterpart polymer solution or dissolution/dispersion of PANI in an appropriate solvent, and polymerization of the matrix in the solution. Despite the fact that, all the described methods of PANI modification, lead to a new class of organic materials with interesting potential application, it is worth emphasizing, that all of them result in materials with lower electrical conductivity as compared to pure, unmodified PANI.

In case of PANI composites, hundreds of matrices have been used to prepare a desirable material. Preparation of PANI/biopolymer composites, being an interesting challenge and has been realized using several matrices that mainly include chitosan [12], cellulose/ cellulose derivatives [13], or gums [14]. On the other hand, starch, which is one of the cheapest and easiest to obtain plant biomaterials, has not been intensively investigated in this field. The only known PANI/starch composites include: chemical polymerization of PANI in aqueous suspension of taro [15] or maize [16] starches which leads to the deposition of PANI at the surface of starch granule.

The idea of obtaining composites comprising a biodegradable polymer, i.e. starch and PANI may enable various applications of these materials. It may result in new materials suitable for application in different industries including: biomedicine, paper industry, food packaging, and many others. Additionally the combination of a natural fully nontoxic, biodegradable material with a synthetic and multifunctional PANI looks as a promising new direction on the field of advanced, functional materials preparation. Starch as an abundant polysaccharide is a cheap and easily available polymer, which can be extracted and purified from natural sources. Therefore, in this study Sago starch was used as a matrix for synthesis of PANI/Sago blends. Sago starch is obtained from sago palm (metroxylon sagu) which typically found in tropical Asia. Sago starch is made of two glucosidic macromolecules; amylose and amylopectin. It has an important reserve polysaccharide which is readily available, low cost, and able to biodegrade. However, the introduction of Sago will reduce the conductivity of the blend drastically. Thus, certain conductive material needs to be introduced to PANI/Sago blend.

Among various conductive materials (such as active carbon, carbon nanotubes, graphene), graphene is the most preferred one. Graphene, a single layer of carbon atoms in a hexagonal lattice, has recently attracted much attention due to its novel electronic and mechanical properties. Therefore, this paper reports the synthesis of Polyaniline/Sago/Graphene (P/S/G) nanocomposites via ultrasonic irradiation technique. The electrical conductivity and morphology of the synthesized nanocomposites will be investigated.

2. Materials and Methods

2.1 Synthesis of P/S/G nanocomposites

Aniline (ANI) (99%) and ammonium persulfate (APS) were purchased from Sigma Aldrich and hydrochloric acid 1 M (HCl) were obtained from TOLSA and deionize water was used to synthesis polyaniline (PANI). Graphene (xGnP® Graphene Nanoplatelets Grade M) was obtained from XG sciences, USA. The PANI nanoparticles were prepared by in situ polymerization of aniline in acidic medium under ultrasonic irradiation in an inert atmosphere. In a 250 mL three-necked flask equipped with inlet and outlet of inert gas, measured quantity of GnP was added to 4 mL (0.0439 mol) of double distilled aniline and sonicated for an hour. Then 100 mL hydrochloric acid (1.0 M) was added; followed by addition of 3 gm of sago into the above solution and kept standstill. Then, the oxidant solution was prepared by dissolving 5 g (0.022 mol) APS in 50 mL water. Both solutions were cooled to 0-4°C in an ice bath. The solutions were degassed with inert gas. The oxidant solution was then added dropwise to aniline salt solution through a dropping funnel under ultrasonic irradiation with a power of 100 W and operated at 40 kHz for 1 h. The obtained solution was kept aside for 24 hrs for reaction to complete. In all reactions, the molar ratio of aniline to ammonium peroxydisulfate was kept at 1:1. In order to remove excess acids and by-products from polymerization, the as-prepared PANI nanoparticles were purified by washing with water three times followed by ethanol until pH 7 was achieved. The dark green powder was collected on a filter paper using vacuum filtration. The filtrate was dried in vacuum (50°C) for 12 hrs.

2.2 Characterization

Electrical conductivity was measured by using four probe Keithley meter (Model 6517 A) coupled with a resistivity chamber (Model 8009) wherein the electrodes are located in a shielded box to minimize stray electrostatic pick-up, which can cause measurement errors. The electrodes are made of stainless steel and are built to ASTM standards. The electrodes in the 8009 resistivity chamber are coated with conductive rubber for better sample electrode contact. Three specimens of each sample were taken for measurement in order to ensure the repeatability and reproducibility of result. For the measurement of electrical conductivity, the powdered conducting PANI nanocomposites of 5 g was prepared in pellet form (diameter: 13 mm, thickness: 1 mm) at pressure of 14 MPa using a Carver model C Press. Morphology of the PANI nanocomposites was studied by using Hitachi HT7700 High Resolution – Transmission Electron Microscope 120kv (HR-TEM 120 KV).

3. Result and Discussion

3.1 Electrical Conductivity:

Electrical conductivity of P/S/G nanocomposites are illustrated in Fig. 1 and reported in Table 1. Four probe method was used for measurement of electrical conductivity of the prepared blends and nanocomposites by pelletizing them (diameter: 13 mm, thickness: 1 mm) at pressure of 14 MPa using a Carver model C Press. The conductivity of neat PANI prepared by ultrasonic irradiation was found to be 1.78 S/cm. As expected upon incorporation of sago starch the conductivity reduced to 3.11786E-4 S/cm. As part of the ongoing research it was found that the higher the polyaniline content in the blend the higher conductivity will be, will is expected and also reported by Zareh et al., [17]. However, since PANI is non-biodegradable, non-flexible and non-processable, only very low amount of PANI is preferred for producing conductive nanocomposites [10]. Therefore, this research focuses on using as low as possible quantity of PANI. The selected blend for incorporation of GnP was P/S-20/80 (20 wt% PANI and 80 wt% Sago).

Table 1. The electrical conductivity of the neat PANI and P/S/G nanocomposites.

Sample ID	GnP (wt.%)	Electrical Conductivity (S cm^{-1})
Neat PANI	-	1.78
P/S	-	3.11786E-4
P/S/G-1	0.125	2.986E-4
P/S/G-2	0.25	0.0014
P/S/G-3	0.5	0.022
P/S/G-4	1.0	0.090
P/S/G-5	1.5	0.138

The electrical conductivity of the HCl doped P/S/G nanocomposites increased with increase in GnP content, as evident from Fig. 1. The increase in the electrical conductivity after the incorporation of GnP was attributed to the additional effects of both PANI and GnP because both are conducting. This shows that there is a π -π interaction between the GnP and the PANI backbone, which leads to a higher conjugated system, as reported by Parveen et al., [18]. This suggests that GnP had a remarkable impact on the microstructure of the resulting P/S/G nanocomposites, which might also lead to an increase in the mobility of the charge carriers in the P/S/G nanocomposites; hence, an increase in electrical conductivity.

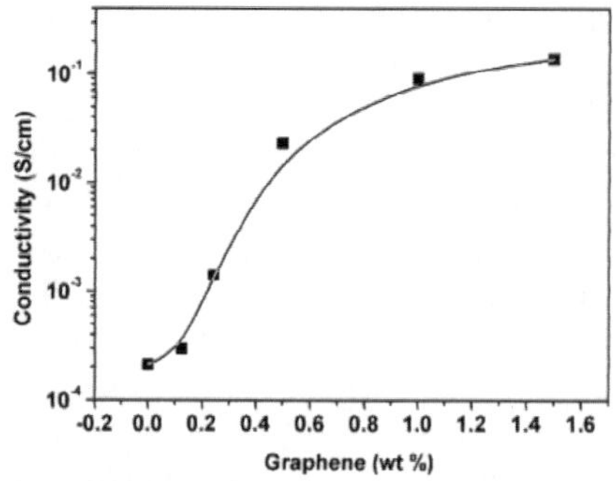

Figure 1. Conductivity value of PANI/Sago/GnP nanocomposites with different compositions of GnP

3.2 Morphology

TEM was performed to examine the inner nanostructures of P/S/G nanocomposites. The TEM image (Fig 2.a) of GnP reveals a sheet like structure with many wrinkles and scrolls [19], which is analogous to the microstructure of the individual GnP sheet. Fig 2.b shows the FE-SEM image of neat PANI, which is tubular in morphology.

Upon synthesis of P/S/G nanocomposites, TEM images shows the presence of both GnP and tubular PANI. The morphology was mostly tubular with certain flakes. The presence of flakes like strucuture is attributed to the adsorption of aniline on the GnP template and its subsequent polymerization. From Fig 3, it can be clearly seen the increase in content of GnP. Fig 3 (a-c) shows uniform dispersion of GnP and also presence of tubular PANI. However, from Fig 3 (d and e) the agglomeration of GnP can be seen. This is further supported by the electrical conductivity values for subsequent nanocomposites. The increase in GnP also resulted in an increase in the diameter of P/S/G nanocomposite compared to pure PANI. The mean diameter of the PANI was measured to be 85 nm, suggesting that they are nanofibers.

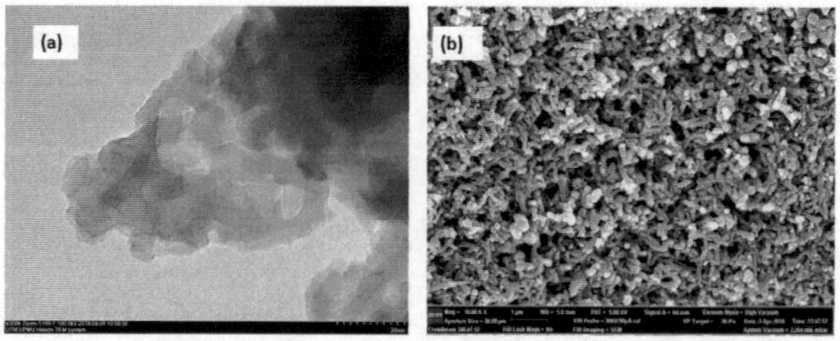

Figure 2. Morphology images of (a) TEM image of GnP dispersed in Sago Matrix, (b) FE-SEM image of Neat PANI.

4. Conclusion

P/S/G nanocomposites were successfully synthesized via ultrasonic irradiation technique. Electrical conductivity results showed that synthesized PANI had electrical conductivity of 1.78 S/cm however upon addition of sago it reduced significantly but at the same time making the PANI/Sago biodegradable, ease of handling and increasing the applications. Further upon introduction GnP to PANI/Sago and synthesising P/S/G nanocomposites the conductivity increased with increase in GnP content and reaching 0.138 S/cm for 1.5 wt% of GnP. TEM analysis confirmed the uniform dispersion of GnP and the adsorption of aniline on the GnP template. Finally, it can be concluded that the prepared nanocomposites are appropriate candidates for different applications requiring conductive nanocomposites.

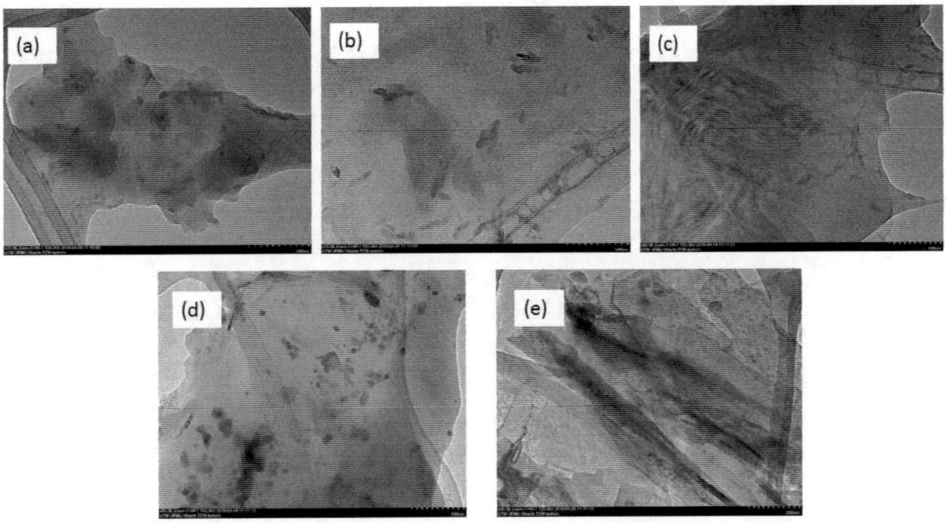

Figure 3. TEM images of P/S/G nanocomposites with varying quantity of GnP (a) 0.125 wt.% GnP (b) 0.25 wt.% GnP (c) 0.5 wt.% GnP (d) 1 wt.% GnP (e) 1.5 wt.% GnP.

Acknowledgements
The authors would like to thank Malaysian government for providing "Malaysian International Scholarship" to M.E. Ali Mohsin, PhD special thanks to IFFCO (M) Sdn. Bhd. for additional financial assistance with contract research grant no. R.J130000.7646.4C141.

References

[1] Molapo KM, Ndangili PM, Ajayi RF, Mbambisa G, Mailu SM, Njomo N, Masikini M, Baker P and Iwuoha EI *Int. J. Electrochem. Sci.* 2012 **7**(12) 11859-11875
[2] Zhao M, Wu XM and Cai CX *J.Phys. Chem. C* 2009 **113**(12) 4987-4996
[3] Li M1, Guo Y, Wei Y, MacDiarmid AG and Lelkes PI *Biomaterials* 2006 **27**(13) 2705-2715
[4] Li D, Huang JX, and Kaner RB *Acc. Chem. Res.* 2009 **42**(1) 135-145
[5] Benhaddad L, Gamby J, Makhloufi L, Pailleret A, Pillier F, Takenouti H *J. Power Sources* 2016 **307** 297-307.
[6] Itoi H, Maki S, Ninomiya T, Hasegawa H, Matsufusa H, Hayashi S, Iwata H and Ohzawa Y. *Nanoscale* 2018 **10**(20) 9760-9772
[7] Mello H and Mulato M *Thin Solid Films* 2018 **656** 14-21
[8] Cichosz S, Masek A and Zaborski M *Polym. Test.* 2018 **67** 342-348
[9] Wan Y, Wu H and Wen DJ *Macromol. Biosci.* 2004 **4**(9) 882-890
[10] Baniasadi H, Ahmad Ramazani SA, Mashayekhan S and Ghaderinezhad F *Synth. Met.* 2014 **196** 199-205
[11] Bhadra S, Khastgir D, Singha NK and Lee JH *Prog. Polym. Sci.* 2009 **34**(8) 783-810
[12] Marcasuzaa P, Reynaud S, Ehrenfeld F, Khoukh A and Desbrieres J *Biomacromolecules* 2010 **11**(6) 1684-1691
[13] Ebrahim SM, Kashyout AB and Soliman MM *J.Polym. Res.* 2007 **14**(5) 423-429
[14] Tiwari, A. and V. Singh. *Carbohydr. Polym.* 2008 **74**(3) 427-434.
[15] Janaki V, Vijayaraghavan K, Oh BT, Lee KJ, Muthuchelian K, Ramasamy AK and Kamala-Kannan S *Carbohydr. Polym.* 2012 **90**(4) 1437-1444
[16] Saikia JP, Banerjee S, Konwar BK and Kumar A *Colloids Surf., B* 2010 **81**(1) 158-164
[17] Zareh EN, Moghadam PN, Azariyan E and Sharifian I *Iran Polym. J.* 2011 **20**(4) 319-328.
[18] Parveen N, Mahato N, Ansari MO and Cho MH *Composites Part B* 2016 **87** 281-290.
[19] Zhou M, Tang J, Cheng Q, Xu G, Cui P and Qin L-C *Chem. Phys. Lett.* 2013 **572** 61-65

IOP Conf. Series: Materials Science and Engineering **522** (2019) 012003 doi:10.1088/1757-899X/522/1/012003

Interface Damage of Protective Layer in TEM Lamella Preparation for Highly Doped Ge Substrate

Nur Nadhirah Mohamad Rashid[1,2], Norhamizah Hazirah Ahmad Junaidi [2] and Siti Rahmah Aid*[1,2]

[1] Malaysia-Japan International Institute of Technology (MJIIT), Universiti Teknologi Malaysia (UTM), Kuala Lumpur 54100, MALAYSIA
[2] MJIIT Microscopy Laboratory, Universiti Teknologi Malaysia (UTM), Kuala Lumpur 54100, MALAYSIA

*Corresponding author (sitirahmah.aid@utm.my)

Abstract. This work reports on the TEM sample preparation of highly doped germanium (Ge) sample using focused ion beam (FIB) technique. The method of the deposition of protective layer is varied to observe the damage at the interface of protective layer and specimen surface. It is shown that TEM lamella prepared using the FIB inevitably contains the surface damage induced by the deposition of protective layer. An effective method of controlling the damage at the interface of protective layer and specimen surface is by introducing a two-step deposition technique of protective layer using electron beam and ion beam, with smooth interface can be observed when the imaging was performed for the cross-sectional TEM.

1. Introduction

Continuous advancement in transmission electron microscope (TEM) has pushed the abilities of the electron microscope technology to even higher resolutions enabling researcher to perform imaging of materials at nanoscale level for specimen's chemistry and crystallography. During the operation of TEM, the specimen need to be uniformly thin enough (< 100 nm) for the high energy beam of electrons to be transmitted and interact with the atoms to gain the aforementioned information. This specimen preparation step is very crucial in TEM and method involved in preparing the sample differs depending on the nature of the material and the information required from it [1,2]. Various preparation methods, such as mechanical polishing techniques and light-ion milling, and focused ion beam (FIB) can be used to obtain cross-sectional TEM specimen. Combination of mechanical polishing techniques and light-ion milling will involve two important steps i.e. grinding/polishing of sample using emery paper (silicon carbide) or diamond lappy film with different grit numbers or particle size, follows by light ion milling using ion slicer or precision ion polishing system. Meanwhile, focused ion beam can directly mill the specimen surface by controlling the energy and intensity of ion beam without polishing the sample. Both ways will eventually result in the TEM specimen with the thickness lower than 100 nm. However, optimization of the specimen preparation steps is important in order to gain the required information from the TEM observation. This work reports on the TEM specimen preparation of highly doped germanium (Ge) sample using FIB technique [3–5]. The method of the deposition of protective layer is varied to observe the damage at the interface of protective layer and specimen surface.

Content from this work may be used under the terms of the Creative Commons Attribution 3.0 licence. Any further distribution of this work must maintain attribution to the author(s) and the title of the work, journal citation and DOI.
Published under licence by IOP Publishing Ltd

IOP Conf. Series: Materials Science and Engineering **522** (2019) 012003 doi:10.1088/1757-899X/522/1/012003

2. TEM Lamella Preparation Method using Dual Beam FIB

In this work, a highly doped Ge (100) substrate, with implanted damaged layer ≈ 60 nm was used for preparing a thin slice of the material, called lamella [3–5]. TEM lamella preparation was demonstrated using dual-beam FIBs machine that enables sample preparation using a focused ion beam of gallium (Ga), and surface imaging of scanning electron microscope using an electron beam, both equipped in two independent microscope columns.

TEM lamella preparation is also known as a lift-out method which consists of five steps:

(a) Deposition of the protective material onto specimen surface on the selected area, follows by the milling of the selected area (cutting pattern) using Ga^+ ions to produce a thin slice of lamella
(b) Extraction of the wedge containing lamella
(c) Lift out of lamella using omniprobe needle
(d) Glue of lamella on top of copper (Cu) grid
(e) Thinning (< 100 nm) and cleaning of lamella using low-voltage focused Ga^+ ion-beam

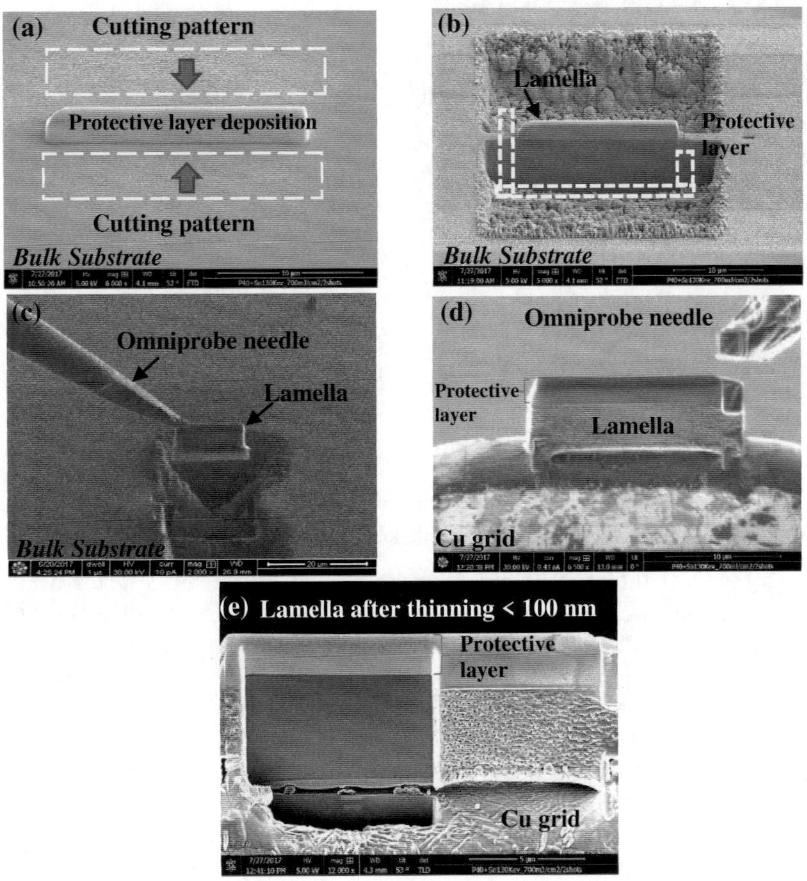

Figure 1. Preparation steps of a FIB TEM lamella

IOP Conf. Series: Materials Science and Engineering **522** (2019) 012003 doi:10.1088/1757-899X/522/1/012003

3. Discussion on the Deposition of Protective Layer

As shown in Fig. 1, the process of FIB lamella preparation starts with deposition of a protective layer on a specific micrometric area on the bulk substrate. This protective layer also functions as a marker to milling specimen for a thin slice of lamella. This protective layer is important in order to provide protection to the top surface of the specimen in the selected lamella area from ion beam induced damage and unwanted milling during the milling and thinning process of lamella. Materials that have been widely used as a protective layer are tungsten (W) and platinum (Pt). Generally, protective layer is deposited by using Ga^+ ion beam with energy of 30 keV, and thickness of 1 μm. We have prepared a lamella sample with a protective layer of W deposited on the substrate using the parameter in the aforementioned technique. The lamella was then prepared by following the steps as mentioned in Fig. 1. The result of cross-sectional TEM of the lamella at the interface area of protective layer and specimen surface is shown in Fig. 2(a).

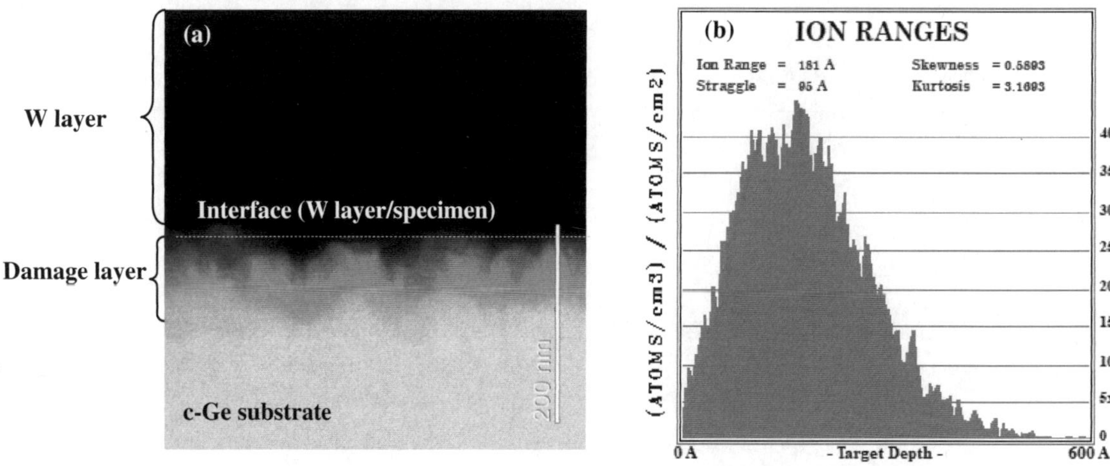

Figure 2. (a) Cross-sectional TEM of lamella prepared by the deposition of W using 30 keV-Ga ion beam (b) Ga ions distribution in Ge substrate at 30-keV ion implantation energy

As can be seen in Fig. 2(a), the deposition of W protective layer using 30-keV ion beam on the surface of highly doped Ge substrate created a rough interface and damage layer with estimated thickness of ≈ 120 nm. This is considered due to the energetics ions beam leads to the inevitable surface damage when the ion is bombarded onto Ge surface. In addition, SRIM-2013 program which calculates the stopping and range of ions into matter using a quantum mechanical treatment of ion-atom collisions [6], shows that Ga ions penetrate deeper into Ge substrate, as illustrated in Fig. 2(b). Fabricated sample is expected to have deeper penetration depth as physical and chemical reactions between the ions and substrate are not taken into account during the calculation [3]. As a result, for the highly doped Ge substrate used in this works, the created interface damage affects the imaging of the original implanted layer which is estimated to be existed ≈ 60 nm near the surface.

In order to control this interface damage, we introduced deposition of protective layer using a low-energy electron beam prior to the ion-beam deposition. We demonstrated this technique by using Pt as the material for the protective layer, with two-step deposition involving 5-keV of electron beam Pt deposition, followed by 30-keV of ion beam Pt deposition with estimated thickness to be ≈ 0.5μm and ≈ 1μm, respectively. Figure 3 shows the deposition of Pt using the aforementioned steps in SEM and FIB images. The total thickness for Pt deposition is estimated to be ≈ 1.5 μm.

The lamella was then prepared by following the steps as mentioned in Fig. 1. The cross-sectional TEM result of the lamella prepared using this two-step technique shows smooth interface between the deposited Pt layer (electron beam) and the top surface of Ge substrate (Fig. 4). It is also shown that damages from the milling and thinning of the lamella only affected the top protective layer of Pt deposited by ion beam, while the layer of Pt deposited by electron beam remain and protect the surface of interest for TEM observation. Implanted damage layer of ≈ 60 nm can clearly be seen from the TEM image, proving the effectiveness of this two-step technique.

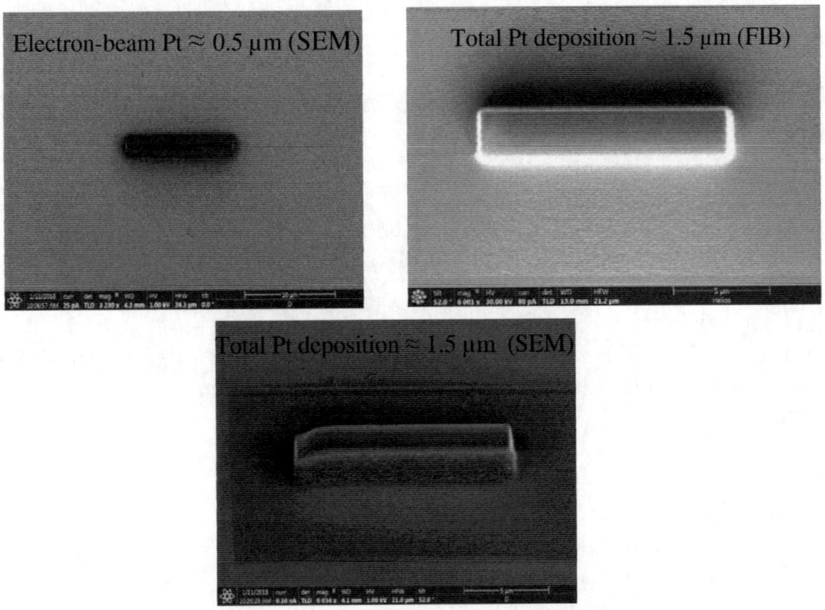

Figure 3. Deposition of Pt protective layer using electron beam followed by ion beam

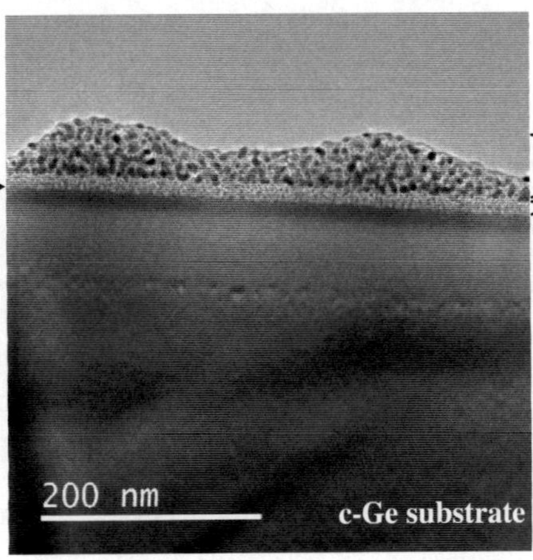

Figure 4. Cross-sectional TEM of lamella prepared by the deposition of Pt using 30 keV-ion beam

4. Conclusion

This work reports on the TEM sample preparation of highly doped germanium (Ge) substrate using FIB technique. Sample preparation steps included are deposition of protective layer, extraction of interest area using milling process, lamella lift-out, glue of lamella on top of copper grid and final thinning process. It is shown that TEM lamella prepared using the FIB inevitably contains the surface

IOP Conf. Series: Materials Science and Engineering **522** (2019) 012003 doi:10.1088/1757-899X/522/1/012003

damage induced by the deposition of protective layer. An effective method of controlling the damage at the interface of protective layer and specimen surface is by introducing a two-step deposition technique of protective layer using electron beam and ion beam. Smooth interface can be observed when the imaging was performed for the cross-sectional TEM. The information gained from this work is important to prepare cross-sectional TEM lamella sample for evaluating specimen such as thin film and shallow pn junction in MOSFET [7 – 11].

Acknowledgments
This work is supported by Takasago Research Grant (OTR) (Vote No: 4B211/4B273/4B314) and Ministry of High Education Malaysia, Universiti Teknologi Malaysia Research University Grant Tier 2 (Vote No: 13J74).

References
[1] Michael Jublot and Michael Texier 2014 *Micron* **56** 63
[2] Miroslava Schaffer et. al. 2012 *Ultramicroscopy* **114** 62
[3] Nur Nadhirah Mohamad Rashid et. al. 2016 *2016 IEEE Industrial Electronics and Applications Conference (IEACon)* 389
[4] Umar Abdul Aziz et. al. 2017 *IOP Conf. Series: Materials Science and Engineering* 204 012003
[5] Nur Nadhirah Mohamad Rashid et. al. 2018 *Ext. Abs. the 17th International Workshop on Junction Technology* **2017** 91
[6] http://www.srim.org/
[7] S R Aid et. al. 2011 *Physica Status Solidi A.* **208** 1646
[8] S R Aid et. al. 2011 *Physica Status Solidi A.* **208** 2772
[9] S R Aid et. al. 2013 *Japanese J. of Appl. Phys.* **52** 026501
[10] S R Aid et. al. 2017 *Physica Status Solidi A.* **214** 1700225
[11] Anthony Centeno et. al. 2018 *Chemosensors* **6(1)** 4

IOP Conf. Series: Materials Science and Engineering **522** (2019) 012004 doi:10.1088/1757-899X/522/1/012004

Influence of anneal temperature in air on surface morphology and photoluminescence of ZnO thin films

Sujun Guan[1], Liang Hao[2], Mariko Murayama[1], Xiaohua Xie[3], Shuji Komuro[4], Xinwei Zhao[1,*]

[1] Department of Physics, Tokyo University of Science, Shinjuku-ku, Tokyo, 162-8601 Japan.
[2] College of Mechanical Engineering, Tianjin University of Science and Technology, Hexi District, Tianjin 300222, China
[3] College of Materials Science and Engineering, Yangtze Normal University, Fuling District, Chongqing 408100 China.
[4] Faculty of Science and Engineering, Toyo University, Kawagoe, Saitama 350-8585 Japan.

* corresponding author's e-mail: xwzhao@rs.kagu.tus.ac.jp

Abstract. The influence of anneal temperature in air ambient under atmospheric pressure on the surface morphology and photoluminescence of ZnO thin films grown on quartz substrate by RF sputtering has been investigated. The characterization of ZnO thin films was carried out by X-ray diffraction (XRD), atomic force microscope (AFM), DRUV-vis spectra (DRUV-vis), photoluminescence (PL), X-ray photoelectron spectroscopy (XPS). XRD analysis shows the crystal quality of ZnO thin films becomes better after annealing in air. AFM results shows that the influence of anneal temperature on surface morphology is obvious. DRUV-vis results reveal that the transmittance increases in the region from 400 nm to 500 nm after annealing, without affecting the band gap. PL spectra of ZnO thin films by annealed at various temperatures consists of a near band edge emission around 380 nm and visible emissions due to the electronic defects, which are related to deep level emissions, which are generated during annealing process. The evolution of defects is analysed by PL spectra based on the energy of the electronic transitions. XPS results hint that the obvious change of oxygen species on the surface of ZnO, especially the increased oxygen vacancies with increasing the anneal temperature.

1. Introduction

Zinc oxide (ZnO) has been well recognized as one of the most promising oxide semiconductor materials owing to its excellent optical, electrical and acoustic properties. Due to a direct band gap of 3.36 eV, ZnO has been found wider applications in recent years, such as light-emitting diodes (LEDs), solar cells, gas sensors, spintronics, transparent conductive contacts (TOCs) [1-3]. Nowadays, ZnO thin film and nanostructures can be fabricated by using many deposition techniques, such as molecular beam epitaxy (MBE), sputtering, pulsed laser deposition (PLD), chemical vapor deposition (CVD), sol-gel technique [4-8]. Among them, radio frequency (RF) magnetron sputtering method has many attractive advantages, which include better film growth control, repeatability, low temperature deposition, large scale stability and uniform film properties [7, 9]. It is well known that the properties of sputtered ZnO thin films are known to depend not only deposition parameters, but also on the post deposition processes, such as thermal anneal. However, the influence of anneal conditions on the ZnO

Content from this work may be used under the terms of the Creative Commons Attribution 3.0 licence. Any further distribution of this work must maintain attribution to the author(s) and the title of the work, journal citation and DOI.
Published under licence by IOP Publishing Ltd

thin film fabricated by RF sputtering are still debatable. Meanwhile, some reports about the effects of the thermal anneal for ZnO thin films focus on the anneal temperature, anneal ambient, such as under a mixed gas under a contain vacuum pressure, or vacuum pressure, and so on [10-15]. However, these approaches require high vacuum and high temperature, and the harsh conditions inevitably result in a high cost. Therefore, it is necessary to develop an alternative method to prepare ZnO thin film with low-cost. In addition, when the anneal temperature up to 400 °C, the resistivity of annealed ZnO thin film becomes high, which would be suitable use as the TCO thin films [15, 16].
In this study, the influence of anneal temperature just in air ambient under atmospheric pressure at 350, 380 400 °C for 1 h, on the crystalline, surface, optical, and photoluminescence behaviour of ZnO thin films is investigated.

2. Experimental

2.1. Fabrication of ZnO thin films

Firstly, the ZnO thin films were deposited on quartz substrate by RF magnetron sputtering. The target is a high purity ZnO (99.99%). The sputtering chamber was initially evacuated down to a pressure of less than 5×10^{-4} Pa. The depositing temperature, working pressure and sputtering power were kept for 30 min at room temperature (RT), 0.5 Pa and 75 W, while the Ar and O_2 flow rates were 0.2 and 29.8 sccm, respectively. Before films growth, the presputtering of the target surface was executed for 5 min to remove possible surface contamination. The ZnO films were deposited on quartz substrate at RT, named as "RT-ZnO". Then, the ZnO films were annealed in air ambient under atmospheric pressure, using muffle furnace (KDF S70) at 350, 380, 400 °C for 1 h, named as "Air-350", "Air-380", "Air-400", respectively.

2.2. Characterization

The crystal structure of the as-prepared samples was analysed by an X-ray diffraction (XRD, Rigaku Ultima IV) equipped with Cu-Kα radiation at 40 kV and 40 mA, employing a scanning rate of 0.01 ° s^{-1}. The surface morphology was examined by atomic force microscope (AFM, Seiko SPA3800N). Photoluminescence emission and excitation were obtained with a fluorometer (PL, iHR320) at room temperature with an excitation wavelength 325 nm. The ultraviolet-visible absorption spectra were measured with a scanning DRUV-vis spectrophotometer (DRUV-vis, Shimadzu SolidSpec-3700/3700DUV). The spectra of O 1s and Zn 2p were analysed by X-ray photoelectron spectroscopy (XPS, Jeol JPS-9010MC).

3. Results and Discussion

3.1. Crystal structure and surface morphology

XRD measurement has been carried out to investigate the crystalline property of the ZnO thin films and annealed ZnO thin films and the results are shown in Fig. 1. The crystal structure of ZnO thin films annealed at anneal temperatures of 350, 380 and 400 °C was identified to be c-axis oriented, which could be attributed to the (002) crystal plane of ZnO (JCPDS # 36-1451), corresponds to a wurtzite ZnO structure. The position of peak has been observed to increase consistently with increasing anneal temperature. The shift in the corresponding diffraction peak position for different anneal temperatures might have been caused due to stress changing in the films [16]. Assuming a homogeneous strain across the films, the crystallite size may be estimated from the full-width at half maximum (FWHM) of (002) diffraction peak using Scherer's formula [17]:

$$D = 0.9\lambda/(A\cos\theta) \quad (1)$$

where λ is X-ray wavelength, A the FWHM of the (002) diffraction peak in radians and θ the Bragg diffraction angle of the (002) peak. The results show that grain size of ZnO (Fig. 1b), is the minimum with a value of ~ 26 nm at RT and maximum with a value of ~ 57 nm at 400 °C. During anneal in air at low temperature, surface atoms had less relative energy, thus the surface mobility of atoms was low, resulting in generation of defects in ZnO thin films. With increasing the anneal temperature, those atoms gained more energy and surface mobility to occupy stable positions inside ZnO crystals. It would help in enhancement of crystalline property of ZnO films [18].

Fig. 2 shows the surface morphology of the samples investigated by AFM measurement. The surface morphology of the RT-ZnO films shows a grain-like structure (Fig. 2a). After annealed at 350 °C, the grain-like structure substantially grows along vertical (Fig. 2b). While increasing the temperature up to 380 and 400 °C, the grain-like structure grows too large then becomes to partially fusion (Fig. 2c and 2d). In other words, with anneal in air, it shows an obvious increased size of the grain. The increased size in the surface morphology is matched with its grain size of ZnO crystals (Fig. 1b).

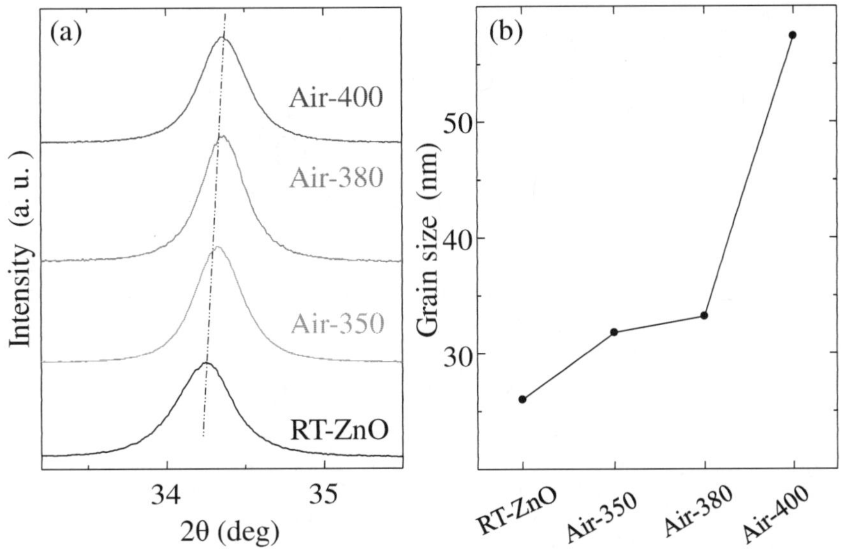

Figure 1. (a) XRD patterns and (b) the grain size of the films.

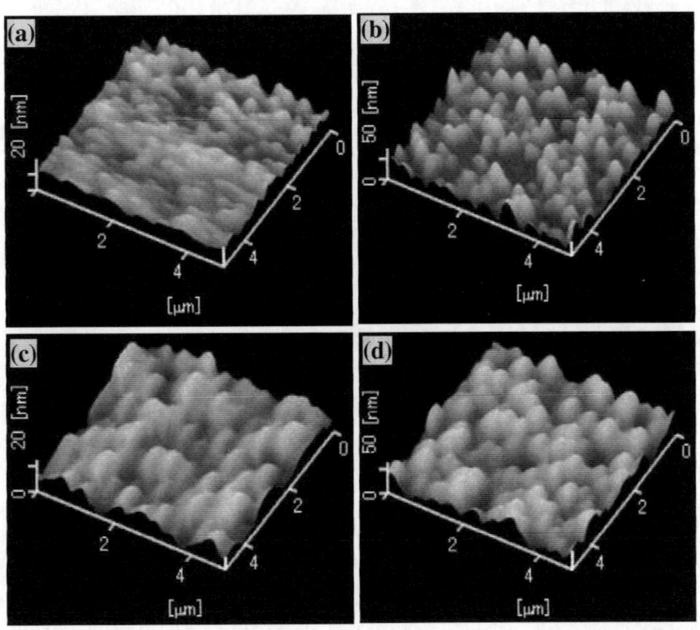

Figure 2. Surface morphology of the samples. (a) ZnO, (b) Air-350, (c) Air-380, (d) Air-400.

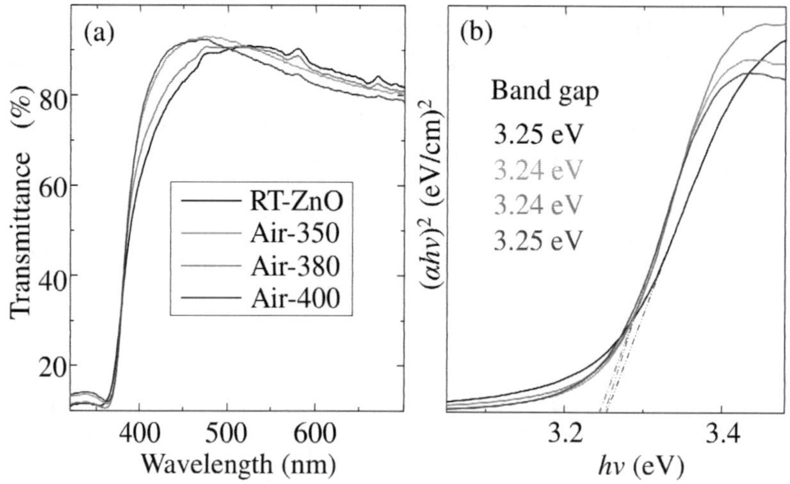

Figure 3. (a) Optical transmittance spectra and (b) plots of $(\alpha h\nu)^2$ against hv of the samples.

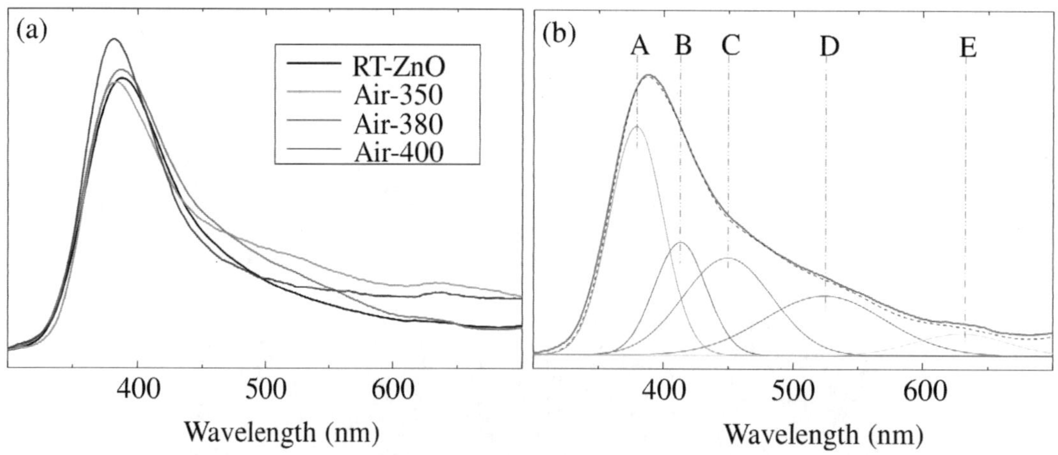

Figure 4. PL spectra of the samples (a) and the fitted curves of Air-380 (b).

3.2. Optical properties

In order to use the TCO thin films in the applications as transparent electrodes for optoelectronic devices, ZnO thin film must be with high transparency in the visible region. DRUV-vis spectrum is used to investigate the light absorption property of the samples. As showed in Fig. 3a, all samples show good optical transmittance. With anneal in air, it is also observed that the transmittance of the annealed ZnO film increases in the region from 400 nm to 500 nm, but the transmittance slightly decreases in the region higher than 500 nm. While the corresponding Kubelka-Munk plots (Fig. 3b), it indicates that the change of band gap of the ZnO film is slight, after annealed in air. The slight change of band gap might be attributed to the formed defect during anneal in air at low temperature [18].

Moreover, the PL emission spectra has been widely used to study the fate of electron and hole in semiconductor [19-23]. Room temperature PL excitation and emission for all samples were measured in the wavelength range of 300-700 nm. PL spectra of ZnO film annealed at various temperatures are shown in Fig. 4a. For RT-ZnO film, the PL emissions show a typical luminescence behavior of a narrow near band edge (NBE) emission in the UV region and defect level (DL) emission in the visible region. The NBE emission could be attributed to the recombination of photoexcitation, while the band emission existing in the visible region is due to the recombination of deep-level holes and electrons [19-23]. With increasing the anneal temperature from 350 to 400 °C, the UV emissions slightly increased, which may be ascribed to the oxidized layers formed on the surface and the grain boundary of ZnO film after annealing in air [24]. While the increase of the visible emissions of the annealed ZnO films could be due to the increase of the intrinsic defects and their concentrations [25].

To further investigate the intrinsic defects, the PL emission spectra of the Air-380 sample has been deconvoluted into five Gaussian components in Fig. 4b. The centre of each peak (A-E peak) is 380, 413, 449, 517, 631 nm, respectively. The peak in ultraviolet (peak A) is due to the recombination of photoexcitation. The peak B appearing through possible transition of electron of electrons trapped at Zn interstitial (Zn_i) defects to the valence band [22, 23]. The peak C is due to possible electron transition between extended states Zn_i (ex-Zn_i) defects level and the valence band [21, 22]. An emission band around 517 nm (peak D) occur through the photogenerated holes trapped in the deep level oxygen vacancy (V_O) recombine with the electrons trapped in a shallow level located just below the conduction band [21, 22]. The peak E could be attributed to possible electron transition between Zn_i defects level and the deep level [22, 23].

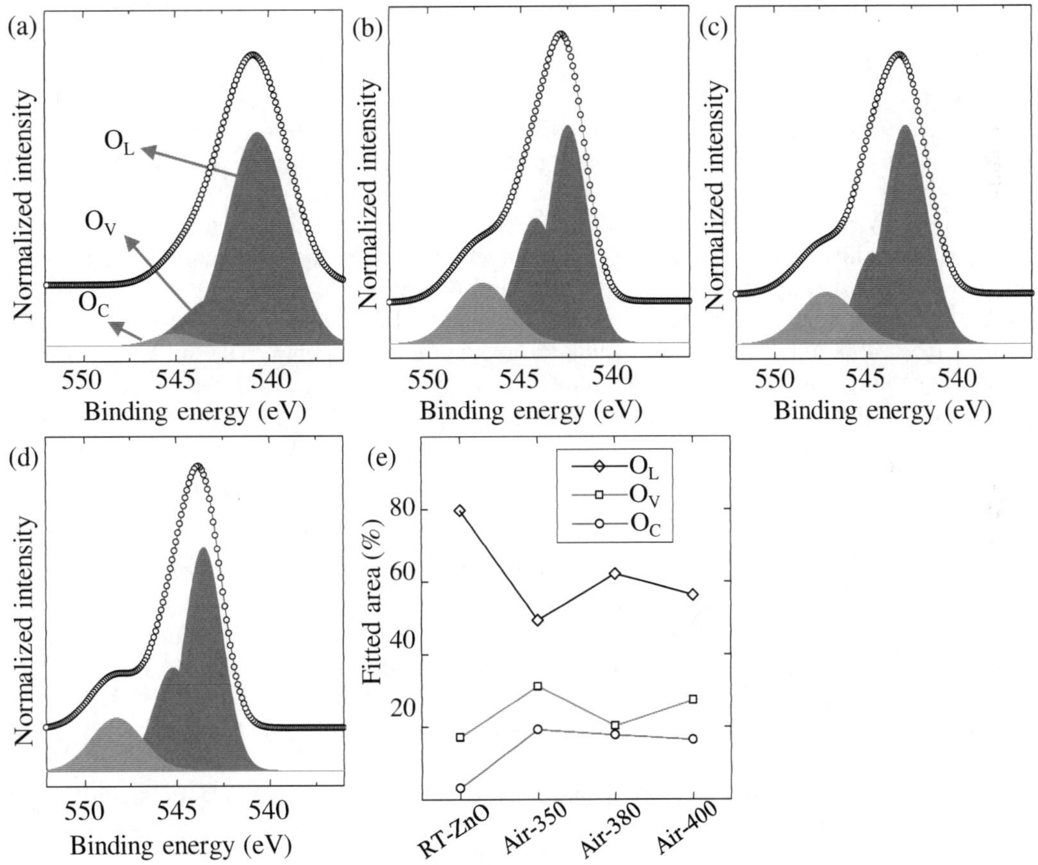

Figure 5. High resolution XPS spectra of O 1s core levels of the samples. (a) RT-ZnO, (b) Air-350, (c) Air-380, (d) Air-400. Peaks are deconvoluted into three components of oxygen, O_L (lattice oxygen), O_V (oxygen vacancy) and O_C (chemisorbed oxygen species), and the relative area ratios of the components evaluated from the fitted curves and plotted in (e).

3.3. Bonding environment

Furthermore, to gain insight into the surface/sub-surface components and oxidation states of the ZnO films has been investigated by with XPS spectra, as shown in Fig. 5. The O 1s peaks of the ZnO films are asymmetric and the deconvolution of the peaks yields three contributions attributable to different chemical states of oxygen using Gaussian fitting. For RT-ZnO film, the lowest peak at 540.6 eV is generally due to the O^{2-} ions, normally surrounded by the Zn ions in the ZnO structure (lattice oxygen, O_L) [26,27]. The higher peaks at 543.0 eV is due to the oxygen-deficient regions, such as oxygen vacancies (O_V). The highest peaks at 545.1 eV is due to the chemisorbed oxygens (O_C) on the film surface [26-28].

Moreover, the results of O 1s peaks of the ZnO films show shift to higher energy with increasing the anneal temperature in air, as shown in Fig. 5a-d. Also, the curve fitting for the O 1s core level peaks is conducted to confirm the relationship between the chemical states of oxygens in the films, and the relative area ratios (%) of the fitted curves are shown as Fig. 5e. Notably, with increasing the anneal temperature in air, it shows that the O_V peaks are enhanced and show highest of the Air-350 sample, which is matched with that of D peak in PL emission spectra. While the O_C peaks are significantly enhanced, which indicates that the loosely chemisorbed oxygen species on the surface increase during annealing in air.

4. Conclusion

In summary, the ZnO thin films were successfully fabricated on quartz substrate by RF magnetron sputtering and subsequent anneal in air ambient under atmospheric pressure. The influence of anneal temperature on crystal structure, surface morphology, optical properties was investigated. The ZnO thin films show a wurtzite ZnO structure. With increasing anneal temperature, the (002) crystal plane of ZnO thin films shift to higher and the grain size increase. The influence of anneal temperature on surface morphology of ZnO thin films is obvious, and the grain-like structure grows too large then becomes to partially fusion. The transmittance of the annealed ZnO thin films obviously increases in the region from 400 nm to 500 nm, without affecting the band gap. The influence of anneal temperature on photoluminescence of ZnO thin films is obvious, especially on the visible range. With the anneal temperature at 350 °C, the annealed ZnO thin films show highest, due to the relative concentration of oxygen vacancies, which is also confirmed by XPS results.

5. References

[1] Fan J, Sreekanth K, Xie Z, Chang S and Rao K 2013 *Prog. Mater. Sci.* **58** 874
[2] Lookk D 2001 *Mater. Sci. Eng. B* **80** 383
[3] Pearton S, Norton D, Ip K, Heo Y and Steiner T 2005 *Prog. Mater. Sci.* **50** 293
[4] Wang H, Liao C, Chueh Y, Lai C, Chen L and Tsiang R 2013 *Opt. Mater. Express.* **3** 237
[5] Gluba M, Nickel N, Hinrichs K and Rappich J 2013 *J. Appl. Phys.* **113** 043502
[6] Chen C, Lan Y, Chou M, Hang D, Yan T, Feng H, Lee C, Chang S and Li C 2012 *Cryst. Growth. Des.* **12** 6208
[7] Chen X, Zhang Z, Yao B, Jiang M, Wang S, Li B, Shan C, Liu L, Zhao D and Shen D 2011 *J. Appl. Phys.* **99** 091908
[8] Bao D, Gu H and Kuang A 1998 *Thin Solid Films* **312** 37
[9] Eisermann S, Sann J, Polity A and Meyer B 2009 *Thin Solid Films* **517** 5805
[10] Chen C, Kelder E, Put P and Schoonman J 1996 *J. Mater. Chem.* **6** 765
[11] Karamdel J, Dee C and Majlis B 2011 *Sains Malaysiana.* **40** 209
[12] Ahn B, Oh S, Lee C, Kim G, Kim H and Lee S 2007 *J. Cryst. Growth* **309** 128
[13] Wang L, Pu Y, Fang W, Dai J, Zheng C, Mo C, Xiong C and Jiang F 2005 *Thin Solid Films* **491** 323
[14] Fang Z, Yan Z, Tan Y, Liu X and Wang Y 2005 *Appl. Surf. Sci.* **241** 303
[15] Nunes P, Fortunato E and Martins R *Thin Solid Films* **383** 277
[16] Pandey S, Pandey S, Awasthi V, Kumar A, Deshpande U, Gupta M and Mukherjee S 2014 *Bull. Mater. Sci.* **37** 983

IOP Conf. Series: Materials Science and Engineering **522** (2019) 012004 doi:10.1088/1757-899X/522/1/012004

[17] Klugg H and Alexander L 1954 *X-Ray Diffraction Procedures* (New York: Wiley)
[18] Kim Y and Kim Y 1997 *J. Vac. Sci. Technol. A* **15** 1103
[19] Wu S, Chan C and Chien C 2016 *Appl. Phys. Lett.***108** 041104
[20] Barman M, Mitra P, Bera R, Das S, Pramanik A and Parta A 2017 *Nanoscale* **9** 6791
[21] Kayaci F, Vempati S, Donmez I, Biyikli N and Uyar T 2014 *Nanoscale* **6** 10224
[22] Das D and Mondal P 2014 *RSC Adv.* **4** 35735
[23] Zeng H, Duan G, Li Y, Yang S, Xu X and Cai W 2010 *Adv. Funct. Mater.* **20** 561
[24] Srinivasan G, Gopalakrishnan N, Yu Y, Kesavamoorthy R and Kumar J 2008 *Superlattice. Microst.* **43** 112
[25] Djurisic A, Leung Y, Tam K, Hsu Y, Ding L, Ge W, Zhong Y, Wong K, Chan W, Tam H, Cheah K, Kwok W and Phillips D 2007 *Nanotechnology* **18** 095702
[26] Chen S, Carraro G, Barreca D, Sapelkin A, Chen W, Huang X, Cheng Q, Zhang F and Binions R 2015 *J. Mater. Chem. A* **3** 13039
[27] Bekermann D, Gasparotto A, Barreca D, Bovo L, Devi A, Fischer R, Lebedev O, Maccato C, Tondello E and Tendeloo G 2010 *Gryst. Growth Des.***10** 2011
[28] Sahai A and Goswami N 2014 *Ceram Int.* **40** 14569

IOP Conf. Series: Materials Science and Engineering **522** (2019) 012004 doi:10.1088/1757-899X/522/1/012004

Authors' background

Name	*Prefix*	*Research Field*	*Email*	*Personal website*
Sujun Guan	Assistant Professor	Transparent solar cell	guansujun@rs.tus.ac.jp	
Liang Hao	Associate Professor	Photocatalysis	haoliang@tust.edu.cn	
Mariko Murayama	PhD Candidate	LED	1218706@ed.tus.ac.jp	
Xiaohua Xie	Associate Professor	Sodium ion battery	xiepingchang@163.com	
Shuji Komuro	Full Professor	LED	skomuro@toyo.jp	
Xinwei Zhao	Full Professor	Nano materials and nano device	xwzhao@rs.kagu.tus.ac.jp	

Note:

[1] **This form helps us to understand your paper better; the form itself will not be published.**

[2] *Prefix*: can be chosen from Master Student, PhD Candidate, Assistant Professor, Lecture, Senior Lecture, Associate Professor, Full Professor

IOP Conf. Series: Materials Science and Engineering **522** (2019) 012005 doi:10.1088/1757-899X/522/1/012005

Numerical modelling of pseudoelastic behaviour of geometrically graded shape memory alloys

Bashir S Shariat, Sam Bakhtiari and Yinong Liu

Department of Mechanical Engineering, The University of Western Australia
35 Stirling Highway, Perth, WA 6009, Australia.

bashir.samsamshariat@uwa.edu.au

Abstract. This study presents numerical modelling based on finite element method for pseudoelastic behaviour of geometrically graded shape memory alloy structures made of near-equiatomic NiTi thin plates. Near-equiatomic NiTi alloy undergoes Lüders-type deformation via stress-induced martensitic transformation over a flat stress plateau with a practically nil stress window over the transformation strain, rendering it impossible for stress controlled actuation. To improve the controllability of a shape memory component in such applications, it is required to have a reasonably wide stress window for the stress-induced transformation in the alloy. One approach to achieving this goal is to design a geometrically graded structure to allow the component to undergo the transformation in different parts under different stress levels. This study investigates the deformation behaviour of geometrically graded NiTi structures with series and parallel design configurations with respect to the loading direction. It reveals that the transformation propagation is along the loading direction in a sample with the series design configuration, whereas it is perpendicular to the loading direction in a sample with the parallel design configuration.

1. Introduction

Owing to their martensitic transformations, shape memory alloys (SMAs) exhibit unique thermomechanical behaviours, most notably the shape memory effect and pseudoelasticity [1, 2]. Shape memory effect is referred to the recovery of a pre-transformation well beyond the elastic limit upon heating. Pseudoelasticity is referred to the recovery of the deformation spontaneously upon unloading. Because of these distinctive features, they have been used in a wide range of applications, such as sensors and actuators [3, 4] and medical appliances [5]. The most widely used SMAs are the NiTi alloys, which undergo a B2→B19′ monoclinic martensitic transformation, where B2 is the austenite (A) and B19′ is the martensite (M) [6]. The transformation can be induced by either the application of a stress or decrease of temperature. When induced by stress, NiTi often undergoes a Lüders-type deformation, which results in a flat stress plateau with almost zero slope over the transformation strain of 6~8% [7, 8]. This phenomenon yields in poor control of displacement by the application of load, which can be a challenge for actuation applications. One approach to solving this problem is to design functionally graded (FG) NiTi to allow the component to undergo the stress-induced transformation in different parts at different stress levels, thus to produce a stress window over the A↔M transformation for better controllability [9-11]. This creates a nonuniform and complex transformation field within the structure.

In FG NiTi, the material properties vary along one or more directions. The property gradient can be acquired in three different ways: by variation in geometry [12-14], in microstructure [15-17], or in

Content from this work may be used under the terms of the Creative Commons Attribution 3.0 licence. Any further distribution of this work must maintain attribution to the author(s) and the title of the work, journal citation and DOI.
Published under licence by IOP Publishing Ltd

composition [18-20]. With respect to the loading axis, the property gradient can be created based on either a series design configuration or a parallel design configuration. In series design configuration, the property gradient is maintained along the loading direction, thus the A→M transformation and the subsequent deformation are expected to occur sequentially along the loading axis [12, 13]. In parallel design configuration, the property gradient is created perpendicular to the loading direction, which yields in transformation propagation vertical to the loading axis. Therefore, the local deformations contribute to the global deformation in a parallel manner [21, 22].

In geometrically graded NiTi structures, the geometrical variation causes the structure to experience a heterogeneous stress field upon loading, thus progressive martensitic transformation within the structure. It has been found that even irregular geometrical variation, such as that in holey or porous structures, can induce nonuniform and complex transformation fields [23, 24]. The concept of geometrically graded SMAs can be applied to the various forms of NiTi alloy, including 1D and 2D engineering components. This article presents finite element modelling of geometrically graded NiTi 2D structures with series and parallel design configurations. The effects of geometrical design on local and global deformation behaviours of such materials are discussed.

2. Finite element analysis and definition of parameters

Finite element analysis (FEA) was conducted using ABAQUS 6.11-2 with a built-in UMAT subroutine based on Auricchio et al.'s constitutive model for SMAs [25, 26]. In this model, the total strain is calculated as the summation of elastic, plastic, and transformation strains. The plastic strain is developed when the recoverable strain limit is exceeded. To simulate the mechanical behaviour of the geometrically graded NiTi samples, the UMAT subroutine has been trained with the actual NiTi alloy behaviour under tension as presented in figure 1. Here, a uniform strip of Ti-50.8at%Ni material was tested at 292 K. The sample had a gauge section of 2 x 30 mm. The forward and reverse transformation stresses were measured to be ~330 MPa and ~80 MPa, respectively. To achieve realistic results by considering all components of normal and shear stresses, a 3D geometry with C3D20R elements has been used for all FEA calculations.

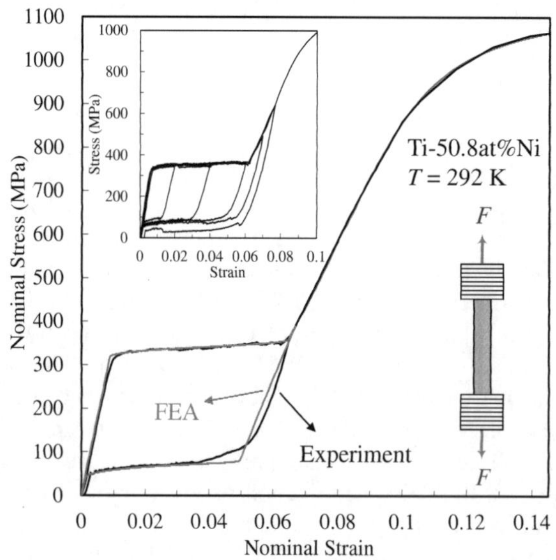

Figure 1. Experimental measurement and FEA prediction of the deformation behaviour of Ti-50.8at%Ni alloy under tension at 292 K.

In this work, the nominal stress (σ) is defined as the load over the initial maximum cross-sectional area of the samples. Also, the nominal strain (ε) is defined as the elongation over the initial maximum length of the samples. For samples with series design configuration, the width aspect ratio is defined as $\alpha = w_1/w_2$, where w_1 and w_2 are the minimum and the maximum widths, respectively as noted in figure 2(a). For samples with parallel design configuration, the length aspect ratio is defined as $\alpha = L_1/L_2$, where L_1 and L_2 are the minimum and the maximum lengths, respectively, as noted in figure 2(b). Also, the taper angle (θ) is defined in figure 2 for series and parallel design configurations. For both design configurations, $\alpha = 1$ indicates a geometrically uniform sample, which exhibits the original material behaviour.

3. Results and discussion

Figure 2 shows the deformation behaviour of the geometrically graded NiTi structures with series and parallel design configurations under tensile loading. The loading direction is along the *x*-axis. Figure 2(a) shows the stress-induced transformation evolution within a series design sample with the width aspect ratio $\alpha = 0.5$, the taper angle $\theta = 88^{\circ}$, and the maximum width $w_2 = 6\text{mm}$. It is seen that the A→M transformation initiates at the top end with the minimum width w_1, where the stress was maximum. Then, the transformation progressively propagates toward the other end with the maximum width w_2 by the increase of the nominal strain ε from 0.01 to 0.07. As observed, the direction of transformation propagation is along the loading direction, i.e. the *x*-axis. Figure 2(b) shows the stress-induced transformation evolution within a parallel design sample with the length aspect ratio $\alpha = 0.5$, the taper angle $\theta = 45^{\circ}$, and the maximum length $L_2 = 30\text{mm}$. As observed, the A→M transformation initiates at the tapered corners on the left side of the sample with the minimum length L_1, where the strain, thus the stress is maximum. Then, the transformation progressively propagates toward the right side of the sample with the maximum length L_2 by the increase of the nominal strain ε from 0.01 to 0.06. It is seen that the direction of transformation propagation is across the lateral direction (along the *y*-axis), which is perpendicular to the loading direction. At 0.06 nominal strain, the entire structure is transformed to martensite, except small areas at the tapered corners on the right side of the sample. It reveals that the acute angle at the corners act as a shield against stress-induced transformation. Figure 2(c) shows the global deformation behaviour of the above two samples over a full tensile loading cycle in comparison with that of a geometrically uniform sample, which corresponds to the original alloy behaviour. As evident, the series design sample exhibits full stress gradient over A↔M transformations. The A→M transformation starts at ~170 MPa (corresponds to the minimum cross-sectional area of the sample) and ends at ~340 MPa (close to the transformation stress of the alloy). The overall transformation strain of this sample is larger than the original transformation strain of the alloy by ~0.014. Similar feature has also been reported in Refs. [12, 13] for geometrically graded SMA structures. The parallel design sample exhibits partial stress gradient over A→M transformation. The A→M transformation starts at ~330 MPa, which corresponds to the actual transformation stress of the alloy. Then, the sample deforms via a stress plateau stage up to ~0.033 nominal strain, where the stress gradient stage begins. During the latter stage, the structure is loaded up to ~630 MPa before unloading. In both design types, the creation of stress gradient over stress-induced transformation provides an expanded stress window over transformation, which enhances the controllability of the NiTi sample in actuation application. The stress window over forward transformation is measured as ~170 MPa and ~300 MPa for the series and parallel design samples, respectively.

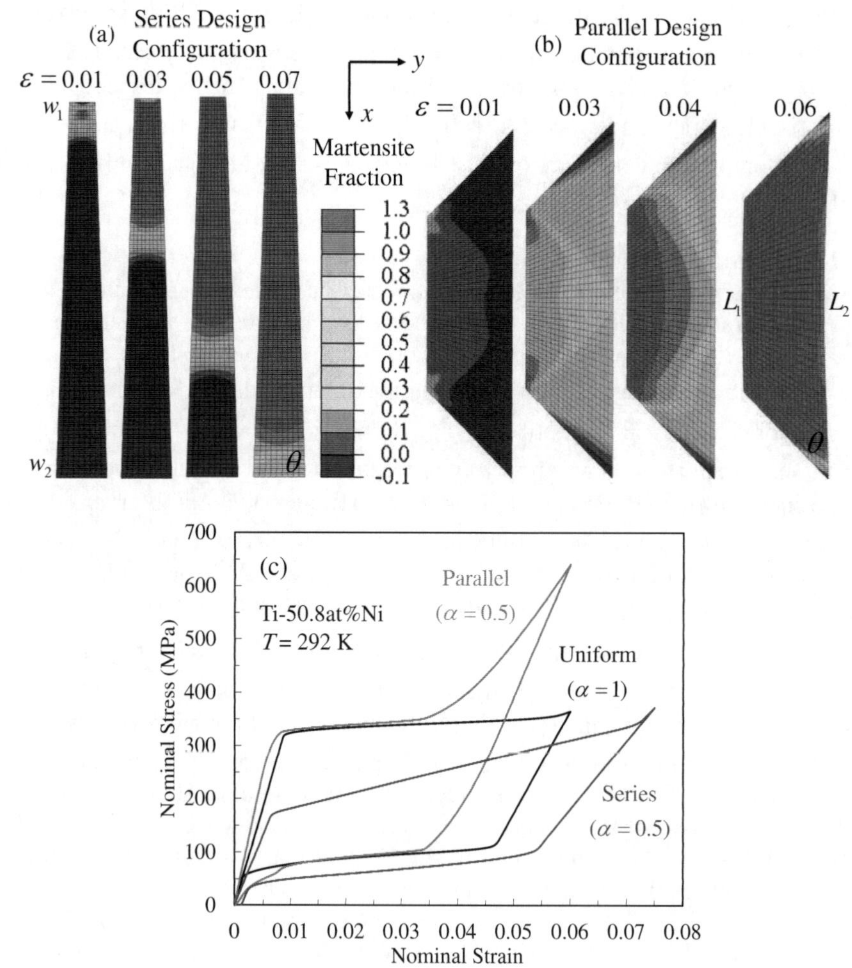

Figure 2. The deformation behaviour of geometrically graded NiTi structures under tension; (a): Transformation propagation in a series design configuration with $\alpha = 0.5$, (b): Transformation propagation in a parallel design configuration with $\alpha = 0.5$, (c): Comparison of the global deformation behaviour of the two design types with that of a geometrically uniform sample.

4. Conclusion

In this paper, the deformation behaviour and the heterogeneous transformation field of geometrically graded NiTi structures were investigated through finite element analysis. The main findings are summarised as below:

(1) The transformation propagation is along the loading direction in a sample with the series design configuration, whereas it is perpendicular to the loading direction in a sample with the parallel design configuration.
(2) Geometrically graded NiTi structures with parallel and series design configurations exhibit partial and full stress gradients, respectively, over martensitic transformation.
(3) The stress gradient over transformation stage provides widened stress window for the geometrically graded NiTi structure, which improves the controllability of the shape memory component. The stress windows of 170 MPa and 300 MPa were achieved, respectively, for the series and parallel design samples with the aspect ratio of 0.5.

Acknowledgments
We wish to acknowledge the financial support to this work from the Australian Research Council in grants DE150101795 and DP180101744.

References

[1] Mohd Jani J, Leary M, Subic A and Gibson M A 2014 *Mater. Design* **56** 1078-113
[2] Shaw J A and Kyriakides S 1995 *J Mech. Phys. Solids* **43** 1243-81
[3] Choudhary N and Kaur D 2016 *Sensors and Actuators A: Physical* **242** 162-81
[4] Huang W 2002 *Mater. Design* **23** 11-9
[5] Morgan N B 2004 *Mater. Sci. Eng. A* **378** 16-23
[6] Bakhtiari S R, Liu J Z, Shariat B S, Yang H and Liu Y 2018 *Scripta Mater.* **151** 57-60
[7] Shariat B S, Liu Y and Rio G 2013 *Mater. Res. Bull.* **48** 5099-104
[8] Shariat B S, Liu Y and Rio G 2014 *Intermetallics* **50** 59-64
[9] Shariat B S, Meng Q, Mahmud A S, Wu Z, Bakhtiari R, Zhang J, Motazedian F, Yang H, Rio G, Nam T-h and Liu Y 2017 *Mater. Design* **124** 225-37
[10] Shariat B S, Meng Q, Mahmud A S, Wu Z, Bakhtiari R, Zhang J, Motazedian F, Yang H, Rio G, Nam T-h and Liu Y 2017 *Data in Brief* **13** 562-8
[11] Shariat B S 2013 Thermomechanical behavior of functionally graded NiTi with complex transformation field, PhD Thesis, The University of Western Australia
[12] Shariat B S, Liu Y and Rio G 2013 *Smart Mater. Struct.* **22** 025030
[13] Shariat B S, Liu Y and Rio G 2013 *Mater. Design* **50** 879-85
[14] Shariat B S, Liu Y and Rio G 2013 *J. Alloys Compd.* **577, Supplement 1** S76-S82
[15] Meng Q, Liu Y, Yang H, Shariat B S and Nam T-h 2012 *Acta Mater.* **60** 1658–68
[16] Shariat B S, Liu Y, Meng Q and Rio G 2013 *Acta Mater.* **61** 3411-21
[17] S. Shariat B, Liu Y and Rio G 2012 *J. Alloys Compd.* **541** 407-14
[18] Cole D P, Jin H, Lu W-Y, Roytburd A L and Bruck H A 2009 *Appl. Phys. Lett.* **94** 193114
[19] Tian H, Schryvers D, Mohanchandra K P, Carman G P and Van Humbeeck J 2009 *Funct. Mater. Lett.* **2** 61-6
[20] Meng Q, Wu Z, Bakhtiari R, Shariat B S, Yang H, Liu Y and Nam T-h 2017 *Scripta Mater.* **127** 84-7
[21] Bakhtiari R, Shariat B S, Motazedian F, Wu Z, Zhang J, Yang H and Liu Y 2017 *Functional Materials Letters* **10** 1740011
[22] Shariat B S, Bakhtiari R and Liu Y 2019 *J. Alloys Compd.* **774** 1260-1266
[23] S. Shariat B, Liu Y and Rio G 2010 *Mater. Sci. Forum* **654-656** 2091-4
[24] Shariat B S, Liu Y and Rio G 2014 *J. Intell. Mater. Syst. Struct.* **25** 1445-55
[25] Auricchio F and Taylor R L 1997 *Comput. Method. Appl. Mech. Eng.* **143** 175-94
[26] Auricchio F, Taylor R L and Lubliner J 1997 *Comput. Method. Appl. Mech. Eng.* **146** 281-312

Characterization of the magnetorheological fluid MRF-140 CG for applications in robotic prosthesis

Oscar Arteaga[1, a], Marcelo Ortiz[1, b], Eduardo Cárdenas[1, c], Segundo M. Espin-Lagos[2, d], Miguel A. Carvajal[1, e], Héctor C. Terán[1, f] and Mario A. Lara[1, g]

[1]Universidad de las Fuerzas Armadas ESPE, Sangolquí, Ecuador
[2]Universidad Técnica de Ambato, Ambato, Ecuador

[a]obarteaga@espe.edu.ec, [b]amortiz8@espe.edu.ec, [c]mecardenas2@espe.edu.ec, [d]sespin@uta.edu.ec, [e]macarvajal@espe.edu.ec, [f]hcteran@espe.edu.ec, [g]malara@espe.edu.ec.

Abstract. The present article shows the different results of the rheological properties of the magneto-rheological fluid (MRF-140CG; LORD Co. (USA), focused on predicting its behavior in devices that use these materials, to characterize the fluid MRF-140CG several tests were performed using a rotational rheometer Anton Paar Physica MRC-501 equipped with a magnetic rheological cell MRD-70/1Text, according to DIN-53018, the tests were performed with magnetic fields from 0 T to a maximum of 800mT with increments of 200mT and temperatures from 20 to 50°C with increments of 10°C in each test. In addition, in order to determine the dynamic model, a shock absorber designed by the authors was tested in accordance with ISO 2954 standards on the Dyno-Shock 11 test bench. To determine which of the models precisely predicts the behavior of the magnetic rheological fluid, a comparison was made between Experimental model, Bingham plastic model, the Bi-Viscose model and the Herschel-Buckley model. This contributed to the design of increasingly sophisticated, reliable and efficient intelligent mechanical devices according to each need.

1. Introduction

Nowadays the technological advance in the field of intelligent materials, among others can be transformed by external agents, such as a magnetic field, which applied to an MRF fluid changes its behavior being non-Newtonian in the presence of the field and Newtonian in the absence of the field, in other words, its state changes from a fluid with low viscosity, to a fluid with high stress with high presence of viscosity. Due to the change in their properties in a controlled way, they are of great interest for applications in engineering fields such as: biomechanical devices, robotic exoskeletons, intelligent prosthesis, etc. [1,2], in suspensions, damping systems, anti-seismic systems, valve control, etc. [3,4] with the purpose of guarantee a technological development innovating with intelligent devices that help to improve people's quality of life.

2. Characterization

In order to characterize the magnetorheological fluids, the parameters are: the type of particle, geometry, size, distribution, and quantity of particles used [5], as far as the carrier fluid is involved, the characteristics of interest are the dynamic behavior of the viscosity and the different additives for the variables of response velocity, deposition and stability. [6] Being necessary for applications with magneto rheological fluids a deep knowledge of their properties and how they are performed at external stimuli to predict the behavior of these parameters and adjust them for application in

Content from this work may be used under the terms of the Creative Commons Attribution 3.0 licence. Any further distribution of this work must maintain attribution to the author(s) and the title of the work, journal citation and DOI.
Published under licence by IOP Publishing Ltd

technological designs that take advantage of these characteristics.; With a focus on these parameters, the characterization of the fluid was carried out by means of two normalized tests, first in a sample, to measure the rheological properties, which are altered under the action of magnetic fields at different intensities and temperatures, followed by a test to determine the dynamic model, which predicts the performance of the fluid, in a magneto-rheological shock absorber, the following models were considered: Empirical model Bingham Plastic model, Bi-Viscose model and the Herschel-Bulkley model.

3. Test apparent viscosity

Realized in rheometer MCR-501, with magnetorheological cell MRD-70/1Text in combination with a measurement system with parallel plates of 20mm diameter of non-magnetizable metal, to avoid radial magnetic forces on the measurement axis, in addition to the thermostatic bath system Julabo F-25 for the application and temperature control with a stability less than ± 0.1°C. Considering the non-Newtonian behavior and the non-constant stress state in the whole sample for the conversion of the parameters of the rheometer rotation angle, rotation speed and torque, and the rheological parameters strain, strain rate and stress the Rabinowitsch conversion was used. [7]

3.1. Test conditions

Fluid tests were performed under the following initial conditions: Temperatures of 20, 30, 40, and 50°C, Heating rate of 0 to 350 rad/s and magnetic field from 0mT to 800mT with 200mT increments in each test.

3.2. Test Results

The processing of the data, as well as their interpretation, is the key to accurately provide all the variables involved in the research, in order to describe the characteristics of the test data, these are processed using descriptive statistical techniques and are presented in graphs of curves to compare and analyze the results.

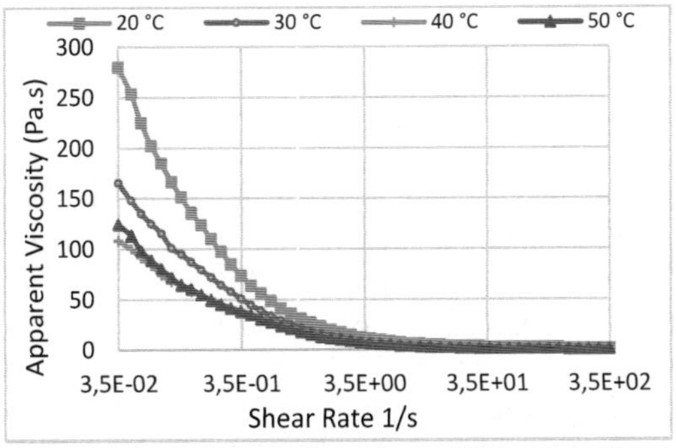

Figure 1. Apparent Viscosity vs Shear Rate 0T.

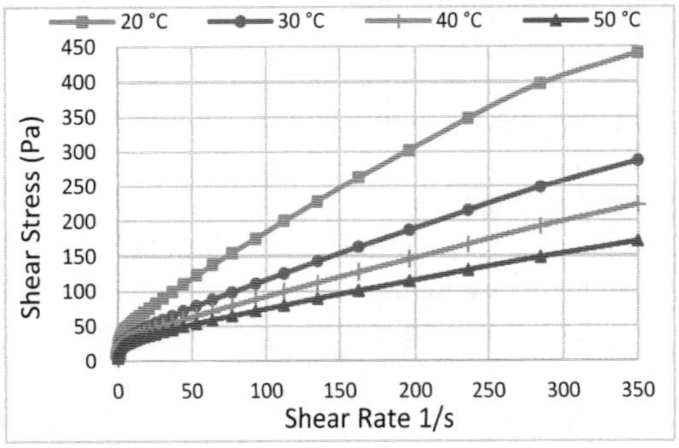

Figure 2. Shear stress vs Shear rate ***0T***.

From the previous results it is concluded that temperature has an effect on viscosity at 0T figure 1 however it does not have an important effect on the apparent viscosity of the fluid MRF-140CG once the magnetic field is applied figure 2. In figures 3,4,5 and 6 the shear stress vs. shear rate is shown for all tests from 0 to 800mT with increments of 200mT.

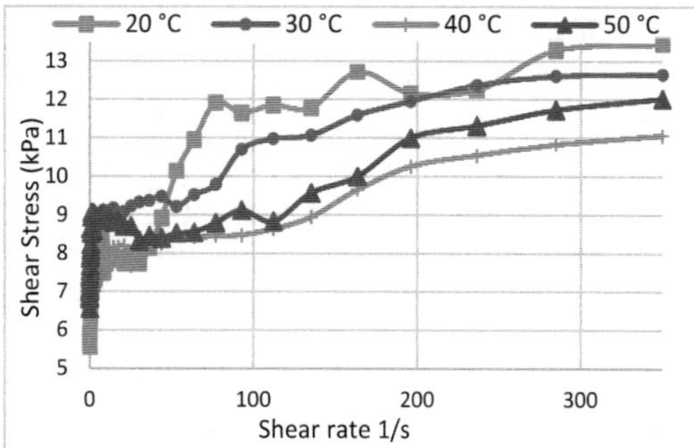

Figure 3. Shear Stress vs Shear Rate with 200 mT.

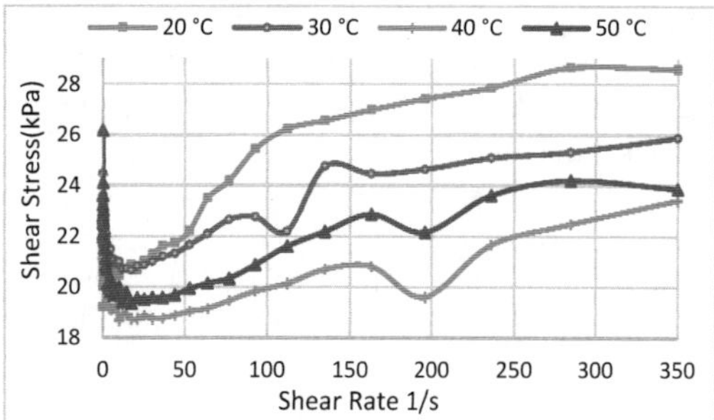

Figure 4. Shear Stress vs Shear Rate with 400 mT.

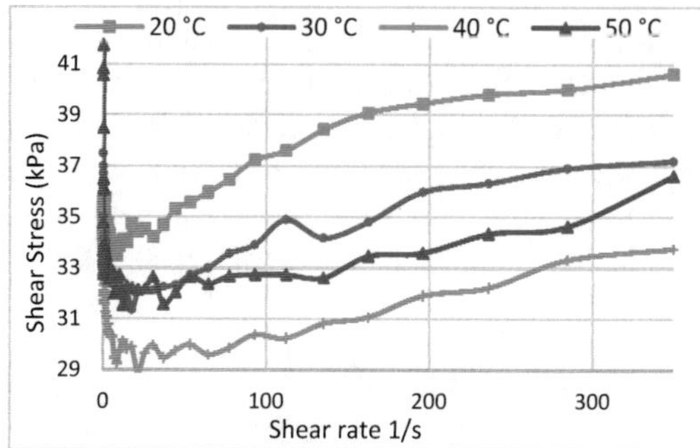

Figure 5. Shear Stress vs Shear Rate with 600 mT.

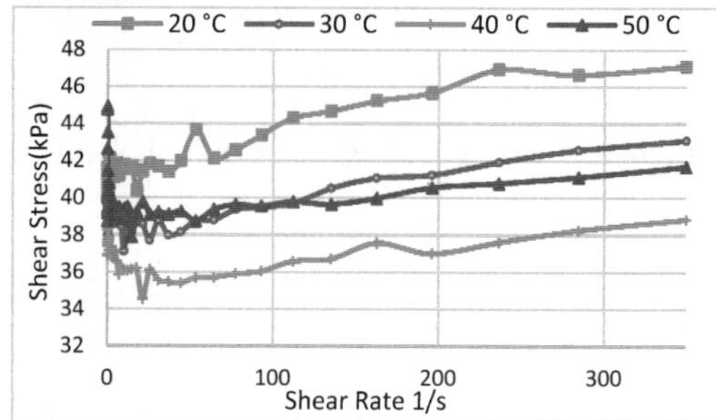

Figure 6. Shear Stress vs Shear Rate with 800 mT.

From these results it is determined that, compared to the off state i.e. in the absence of magnetic field the viscosity increases a percentage of 65631.74% with a field of 200mT figure 3, an increase of 228.26% for 400mT figure 4. then with 600mT it decreases to 58.63% figure 5, and finally at 800mT it falls to 14.02% figure 6, i.e. the fluid begins to saturate until it reaches full saturation at approximately 1.7T, from which the apparent viscosity remains constant.

4. Test comportment dynamic model

Different theoretical models have been proposed to predict the dynamic behavior of the MRF and each one of these can be more specific than the other depending on the application and the type of device in which it is wanted to use. In the case of robotic and exoskeletal prostheses the loads are small compared to applications in suspensions and automotive shock absorbers for this reason, it is important to select the model that best adapts to our specific application.

In order to validate the accuracy of each model, a total of 10 laboratory tests were carried out on the magneto-rheological shock absorber, according to ISO 2954, on the Dyno-Shock 11 test bench, under the following conditions: Displacement of 40mm with frequencies of 1.59; 3.18; 4.77 and 6.37 Hz and speeds of 125; 200; 400; 600 and 800 mm/s at a temperature of 25°C with a heating speed of 200mm/s.

4.1. Empirical model.

For the empirical model by means of equation (1), can obtain the magnitude of the total resistance to yield [8],

$$\tau_{T(H)} = C\ 271.700\ \emptyset^{1,5239} tanh(6,33 \cdot 10^6\ H) \quad (1)$$

Where:

$\tau_{T(H)}$= Magnitude of total resistance yield
H = External magnetic field

4.2. Bingham Model

For the Bingham plastic model, it is considered that the plastic viscosity of the fluid remains approximately constant and the shear stress that must be exceeded to initiate the flow is equal to total resistance yield. Is given by equation (2)[9].

$$\tau_{T(H)} = \tau_{y(H)} + \eta_p \dot{\gamma} \tag{2}$$

Where:

$\tau_{y(H)}$ = Shear stress of the fluid controlled by the magnetic field and defined by the intercept point of the linear regression line.

η_p = Plastic viscosity obtained from the slope of the linear regression line of the shear stress as a function of the Shear rate.

$\dot{\gamma}$ = Shear Rate.

4.3. Bi-viscous Model

For the Bi-viscous model, especially in devices working in squeeze mode, equation (3) and equation (4) govern the total resistance yield.

$$\tau_{T(H)} = \begin{cases} \tau_{y(H)} + \eta_p \dot{\gamma} & |\tau| > \tau_1 \\ \eta_r \dot{\gamma} & |\tau| < \tau_1 \end{cases} \tag{3}$$

$$\tau_{y(H)} = \tau_1 \left(1 - \frac{\eta_p}{\eta_r}\right) \tag{4}$$

Where:

η_r y η_p = Viscosities related to the properties of the fluid in the elastic zone and in the viscous zone respectively.

τ_1 = Shear stress of fluid before yield.

4.4. Herschel-Bulkley Model

For the Herschel-Bulkley model, in which the fluid is subjected to thickening or thinning by cutting after the yield point, particularly when subjected to high shear rate, the model is expressed in the equation (5) [10].

$$\tau_{T(H)} = \tau_{y(H)} + K\dot{\gamma}^n \tag{5}$$

Where:

K = Consistency factor

n = Index of flow behavior and is defined by the exponent of the potential trend curve, for $n >$ 1, represents a cutting thinning fluid, cutting thickening fluids are described with $n > 1$, for $n =$ 1, the model is reduced to Bingham Plastic Model. Si $\tau_T > \tau_{y(H)}$ the fluid behaves like a fluid, otherwise like a solid.

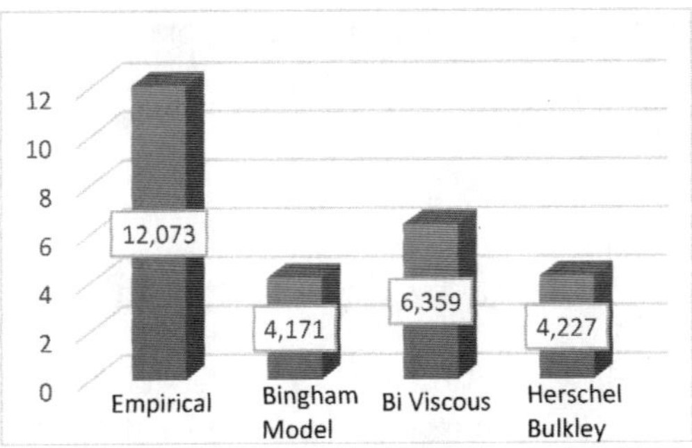

Figure 7. Average error ratio of each model

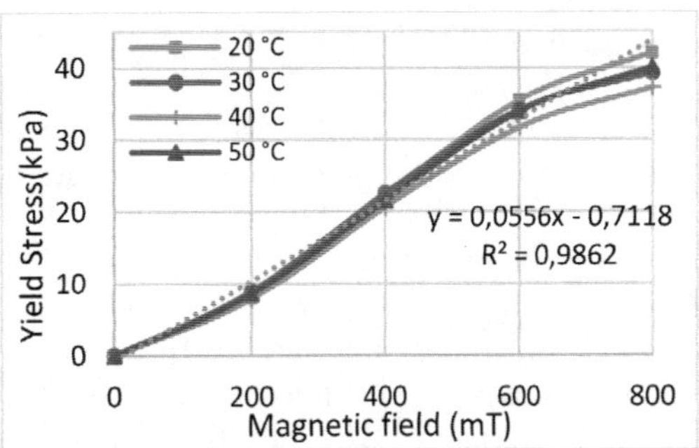

Figure 8. Shear yield stress of MRF-140CG as a function of the magnetic field.

IOP Conf. Series: Materials Science and Engineering **522** (2019) 012006 doi:10.1088/1757-899X/522/1/012006

4.5. Test Results

Based on these results from the four models analyzed, the percentage of average error is determined for each one of them figure 7, because Bingham's plastic model has the lowest average error is the best one to predict the behavior of MRF-140CG fluids, this model is used to find the equation of flow resistance as a function of a magnetic field of specific intensity. Then the results are plotted to obtain the regression equation that defines the relationship between the flow resistance and the magnetic field. In Figure 8, it can be appreciated that the yield stress increases as the magnetic field stress increases. By means of linear regression the equation 6 of the curve is obtained, with the square of the relation of person equal to 0,9862 which indicates a positive correlation.

$$y = 0{,}0556x - 0{,}7118 \tag{6}$$

Where:

y magnitude of total resistance yield ($\tau_{T(H)}$) en kPa.

x magnetic flux density (B) en mT.

5. Conclusions

Bingham's plastic model is the one that most accurately predicts the dynamic behavior of the MRF-140CG fluid under the influence of a magnetic field, with an error of only 4.171%. Therefore, it is safe to use this model for the design of any device that optimally exploits the mechanical properties of the magneto-rheological fluid studied.

For magnetic fields below 200mT, the flow resistance of the fluid is linearly proportional to the field intensity and reaches a value of 8,035 kPa. However, for magnetic fields greater than 200mT, a shear stress of 54,896 kPa is obtained in the case of the design of a prosthesis with a magnetic field of 1T by means of equation 6.

References

[1] O Arteaga, D Camacho, S M Espín, M I Erazo, V H Andaluz, M Mounir Bou-Ali, J Berasategi, A Velasco and E Mera 2017 Characterization of magnetorheological fluids applied to prosthesis for lower limbs active damping *Lecture Notes in Computer Science* **449** 239-247

[2] M Ashtiani, S H Hashemabadi and A Ghaffari 2015 A review on the magnetorheological fluid preparation and stabilization *J. Magn. Magn. Mater.* **374** 716-730

[3] Ubaidillah, A N S Permata, S A Mazlan, D Tjahjana and P J Widolo 2017 There-dimensional finite element magnetic simulation of an innovative multi-coiled magnetorheological brake *IOP Conf. Ser. Mater. Sci. Eng.* **257** 012052

[4] Y Shiao and Q Nguyen 2013 Development of a multi-pole magnetorheological brake *Smart Mater. Struct.* **22** 065008

[5] T Kikuchi, I Abe, A INOUE, A Iwasaki and K Okada 2016 Characteristics of a magnetorheological fluid in high shear rate *Smart Mater. Struct.* **25** 115021

[6] C Molina, M Ardilla, A Hernández 2014 Modelling of Magnetorhelogical Fluid Produced by Mineral Magnetite *Ingenio Magno* **5** 120-7

[7] J Berasategui, M J Elejabarrieta and M M Bou-Ali 2014 Characterization analysis of a MR damper Smart Mater. Struct. 23 01-12

[8] J D Carlson 2005 MR fluids and devices in the real-world *Int. J. Mod. Phys.* **19** 1463-70

[9] D H Wang and W H Liao 2011 Magnetorheological fluid dampers: a review of parametric modelling *Smart Mater. Struct.* **20** 01-34

[10] G Yang, B F Spencer, J D Carlson and M K Sain 2002 Large-scale MR fluid dampers: modelling and dynamic performance considerations *Eng. Struct.* **20** 309-323

IOP Conf. Series: Materials Science and Engineering **522** (2019) 012007 doi:10.1088/1757-899X/522/1/012007

Optimization of Bilayer Actuator Based on Carbon Black/Polymer Composites

Shivashankar H [1,*], Sangamesh R [2], and S M Kulkarni[1]
[1]Department of Mechanical Engineering, National Institute of Technology, Surathkal, Mangalore 575025, India
[2]Department of Metallurgical and Materials Engineering, National Institute of Technology, Surathkal, Mangalore 575025, India

E-mail: *sssnitk@gmail.com

Abstract. In the last few years, actuators based on polymer composite have been created for incredible potential applications in the zone of artificial muscle, micro-robots, relays, and energy harvesting. Polymer composites show the more massive deflection or bending due to the electrothermal and photothermal efforts. Subsequently, these have excellent orientation on the effect because of material properties and structure. In this study, theoretical modeling is employed to understand and analyze the actuator performance by incorporating carbon black (CB) into the polymer material. Polydimethylsiloxane (PDMS) acts as a polymer matrix with bilayer geometry. The displacement of bilayer polymer composite is identified by the length and thickness of two layers, the distinction of coefficients of thermal expansion (CTE) between bilayer and temperature change are inspected. Theoretical outcome demonstrates that the displacement is enormously affected by the thickness proportion of bilayer actuator. In this manner, it is optimized by upgrading thickness proportion and distinct parameters of the bilayer actuator. Thus, this investigation will give a hypothetical reference to the realistic design and realization of the CB /PDMS composite based on a thermal input.

1. Introduction

There is growing interest in the design of the actuator system by mimicking nature to yield lifelike motion in response to external stimuli, such as heat, electrical, light, and moisture [1,2]. In recent year, researcher have been exploring materials, which can provide large deformations to different stimuli. Piezoelectric and magnetostrictive materials deliver smaller deflections and consume more power. Shape memory materials are able to produce more considerable deformation but have relatively longer response time [3-5].

Presently many researchers have been interested in polymer composite kind of actuator study. In 2008, Hai et al. have developed a computational model of single-wall carbon nanotube polymer composite for thermal conductivity study [6]. In 2014, Ying et al. have designed and fabricated graphene and polymer composite bimorph actuator has a large deflection under low voltage (1.2 cm-1 at 10V for 3s) [7]. Thus, comparing with conventional material, polymer materials exhibit high performance, lightweight, fast response, and ease of fabrication. In this way, for the actuator development, polymer composite based materials are widely explored in recent year [8-10]. Polymer-based composite materials possess good electrical, thermal and mechanical properties [11, 12]. Carbon black is introduced into polymer base material to formulate composite material; CB material will enhance the features and performance of the base material [13]. Thus, carbon black placed polymer

Content from this work may be used under the terms of the Creative Commons Attribution 3.0 licence. Any further distribution of this work must maintain attribution to the author(s) and the title of the work, journal citation and DOI.
Published under licence by IOP Publishing Ltd

IOP Conf. Series: Materials Science and Engineering **522** (2019) 012007 doi:10.1088/1757-899X/522/1/012007

composite is used in the micro gripper, micro-robots, micro heater, energy harvesting, sensors, and actuator so on. The composites can transfer different stimulus input into heat and uniformly distributed on to polymer composite. Thus, the thermal properties of carbon black filled polymer composite material will vary and diverted into deflection [14].

The main principle of a thermally induced actuator is adopting two layers of material, which has a different coefficient of thermal expansion between the two materials and forming a bilayer geometry. Hence due to a difference of CTE, the internal stress of the bilayers makes the actuator bends away from the larger CTE. Thus, there are few reports on the numerical analysis of thermally induced deflection based on carbon black and polymer composite. In this paper, a hypothetical examination of bilayer actuator utilizing PDMS and CB-PDMS composite is done. Relationship of a certain important parameter such as geometry, CTE, temperature change with respect to the displacement of bilayer composite actuator is studied. This optimization work will be further extended for appropriate design and fabrication of CB/polymer composite based thermal actuator.

2. Material and Geometry

The geometry of the bilayer strip consists of pristine and composite materials actuator is appeared in figure.1. The material introduced for the analysis of thermal actuator is carbon black and polydimethylsiloxane. PDMS is an organic silicone polymer compound, which is frequently called silicone elastic. It is also known as optical transparency elastomer cured at room temperature silicon rubber. At last cured PDMS has exceptional thermal and mechanical properties [15].

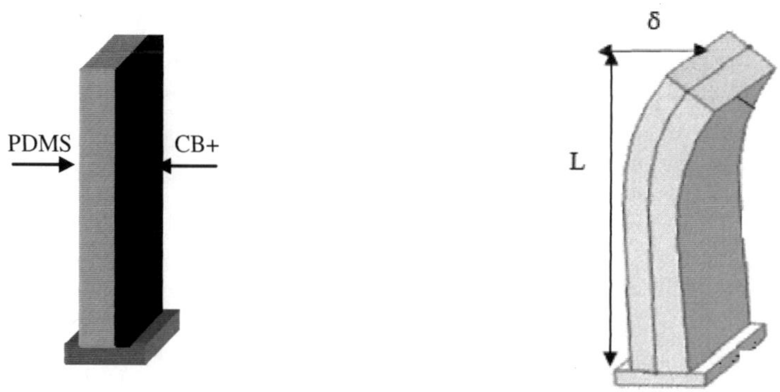

Figure 1. The geometry of a bilayer actuator.

Thus, the polymer composite material is a double layer structure. One layer is composite of CB and PDMS material, is also named as CB + layer, since, the content of carbon black is more and in other layer named as pristine PDMS layer. Three-dimensional size of the bilayer polymer actuator is included here is 20mm x 5mm x 0.50mm (length x width x thickness). The properties of PDMS and CB+ are shown in Table 1. The Young's modulus, CTE of the polymer composite parameters obtained according to an inverse law by adopting the equation 1 and equation 2.

$$E_c = \frac{1}{\frac{v_f}{E_f} + \frac{v_m}{E_m}} \quad (1)$$

$$\alpha_c = \frac{1}{\frac{v_f}{\alpha_f} + \frac{v_m}{\alpha_m}} \quad (2)$$

Table 1. The material parameter used for analysis.

Material	Young's Modulus(E)[MPa]	Density (ρ) [kg/m^3]	Poison's ratio (v)	The coefficient of thermal expansion (α) [1/^{0}C]
Pristine PDMS	0.85	1030	0.45	3.10x10^{-4}
CB/PDMS composite	1.57	1054	0.43	1.16x10^{-4}

Where, m, f and c represented as matrix, filler, and composite of the material. E_c and α_c speak to Young's modulus and CTE of the CB+ composite layer, E_f and α_f are a reference as Young's modulus and CTE of the carbon black filler material. Furthermore, E_m and α_m are Young's modulus and CTE of the pristine PDMS material. Communicated v_f, v_m as the volume fraction of carbon black material and PDMS material. The estimation of the filler and matrix content is figured to 0.05% and 0.95% individually.

When heating of the bilayer composite actuator, because of a distinction of CTE, the deformation is far from the source, due to pristine PDMS layer has a bigger CTE than the CB+ layer. The solid holding of the two layers brings about the entire actuator creating internal stress and bending to towards the CB+ layer. The boundaries are assumed that one of the ends is settled and another end has free displacement. While stop heating, bilayer will cool off, and the actuator comes back to the original position.

3. Theoretical Modeling

The schematic illustration of the bilayer actuator bending is as appeared in figure 2. In the theoretical examination, some of the assumptions are proposed for bilayer deformation. An assumption is made like (i) There is a uniform heat conveyance inside the actuator and disregarding the heat misfortune (ii) The CTE of material does not change with temperature (iii) The bilayer actuator has a direct appropriation of stress, keeping away from the boundary impacts of the actuator. The displacement of the bilayer actuator is affected by CTE, Young's modulus and temperature change of the bilayer. The horizontal displacement (δ) of the free end of the bilayer cantilever can be expressed with the following relationship.

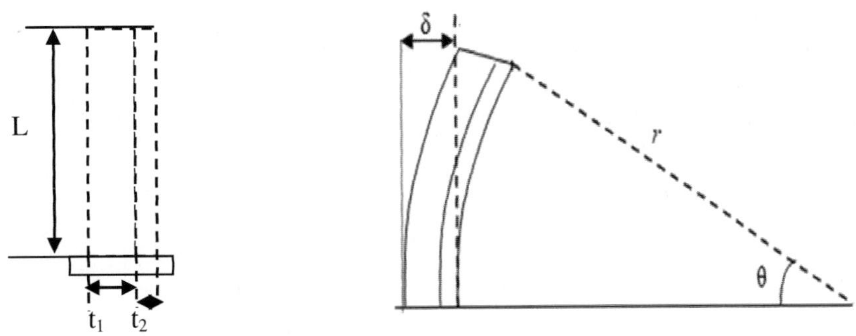

Figure 2. Schematic diagram of a bilayer actuator.

$$\delta = r - r\cos(\theta) \quad (3)$$

Where r and θ are the radius of the curvature and deflected angle appears in the above figure. Whereas, θ=L/r, L is the length of the actuator. The radius of the curvature (r) of the bilayer actuator can be communicated as [16].

$$\frac{1}{r}=\frac{6W_1W_2E_1E_2t_1t_2(t_1+t_2)\,\Delta\alpha\Delta T}{(W_1E_1t_1{}^2)^2+(W_2E_2t_2{}^2)^2+2(W_1W_2E_1E_2t_1t_2)(2t_1^2+3t_1t_2+2t_2^2)} \qquad (4)$$

Where, W, t, E, and α are the width, thickness, Young's modulus, and CTE of bilayer actuator of pristine PDMS (with subscription 1) layer and CB+ (with subscription 2) layer respectively, and ΔT is the temperature rise from the room temperature, where Δα is given by the difference in coefficient of thermal expansion (α_1-α_2) of respective material. It is observed from equation 3 and equation 4 of the free end displacement of bilayer actuator, Is mainly depends on length, Young's modulus, thickness, and CTE of the layers. Hence, the optimization of the bilayer actuator decided by many factors. In that, some of the more influencing parameters explored such as ratios of the thickness, length of the actuator, difference in CTE and temperature change.

4. Results and Discussion

In case one, the proportion of the thickness is considered for examination, where the thickness of CB+ layer is set as a fixed value and other layer PDMS is considered as a variable. In this study, the thickness variation of the bilayer actuator impacts the high deformation. Subsequently, the factor's is represented as the thickness proportion of the two layers, given by $m=t_1/t_2$; where t_2 is approximately fixed to 20 µm and after that $t_1=m \times t_2$ obtained by changing other layer thickness. Alternate parameters utilized for estimation appeared in Table 1. With fixed temperature is substituted in equation.4 to get the curvature of the bilayer actuator.

The examination result appeared in figure 3. The free end of the bilayer actuator displacement is quick increments with the expansion in thickness proportion. It achieves the highest estimation of approximate 11.5 mm at the thickness proportion around 10. Afterward, the displacement slowly decreases with increases in thickness ratio.

Along these lines, the proportion of the thickness is plying the vital part in design and optimization. Since, by considering the thickness of the perfect PDMS layer is thin, for example, near zero, at that point the proportion of thickness winds up zero. Then, the actuator bilayer became a single layer structure of the CB+ layer. Subsequently, there is no distinction between the thermal expansion of the layer. Accordingly, the displacement tends to be invalid after heat input. For another situation, if the thickness of the pure PDMS layer is thick and actuator progressed toward becoming bilayer structure, which has a sizeable substantial contrast in thermal expansion, still the displacement tending to be invalid also. Since the internal stress is exceptionally hard to make such thick actuator deflect after heat input. In a streamlined word, the deflection tends to zero, while the thickness proportion has a tendency to be zero and greatest respectively. Thus, there must be one optimum thickness ratio, where the displacement is highest. In this theory, the assumptions were made for investigation is affirmed that to achieve more noteworthy deflection for sensible application, Hence, need to select an optimum thickness ratio of the bilayer composition.

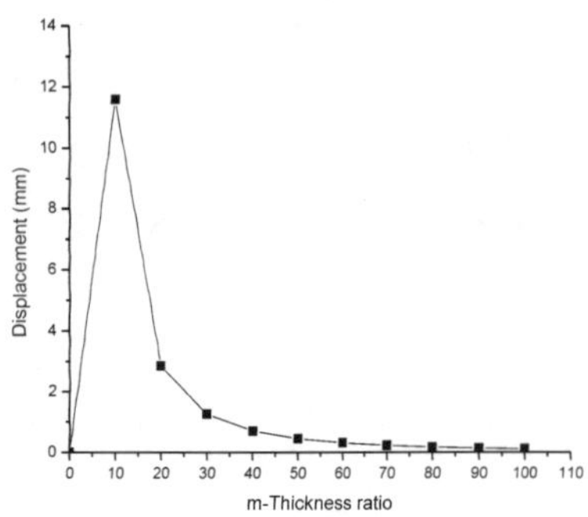

Figure 3. Displacement due to a ratio of the thickness.

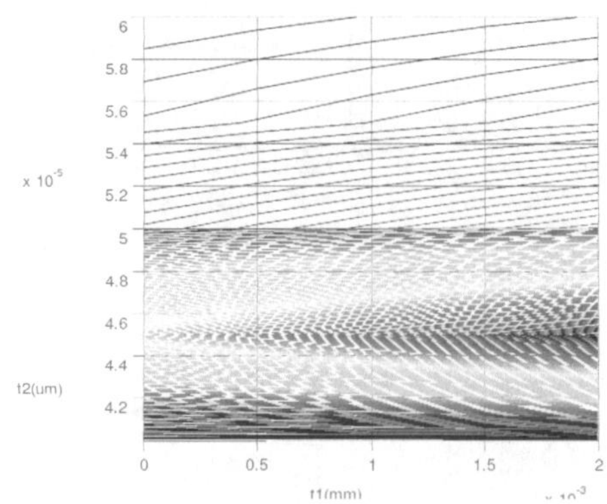

Figure 4. Contour surface map of variable thickness.

In case two, both the thickness of the bilayer fluctuates. The diversion of the bilayer observed by varying both the thickness of CB+ and pure PDMS layer by fixed input temperature. The analytical outcomes appear in figure 4 and figure 5.

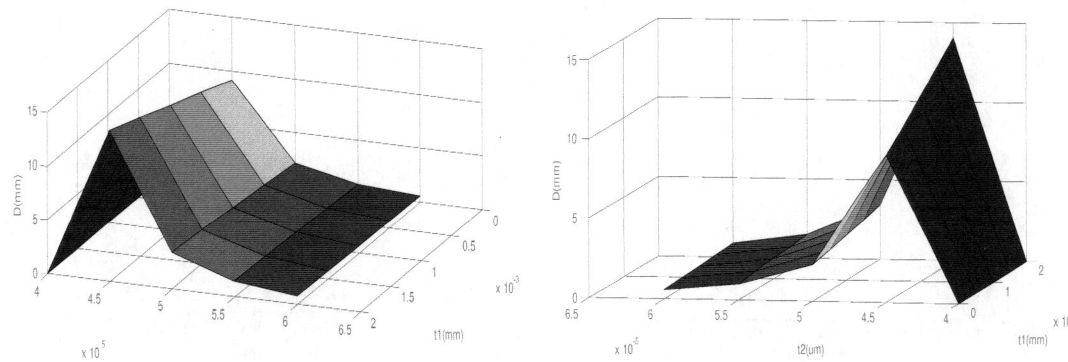

Figure 5. Demonstrates the three-dimensional results of thickness dependance displacement

The three-dimensional results demonstrate that, the actuator of the displacement (D) reliance on the thickness of the bilayer. The contour map depicts that area of the thickness proportion around 33 to 35, where the deformation is moderately high. Keeping in mind the end goal to get more prominent displacement, the thickness of the layers ought not to be too large or too small. From this investigation recommends that pure PDMS layer range is 1-1.5 mm and CB+ layer around 43-46 μm.

In case third, actuator deflection is likewise impacted by the length of the beam. The main key parameter for micro-actuation is the ratio of deflection to length (δ/L). This proportion enables the actuator to bend more with the fixed thickness. As in figure 6 (a) predict the analytical result of the deflection to length proportion relies upon thickness proportion and the variation of effect illustrate as same as a figure. 3. Also, the free end displacement of the actuator is expanding with the expansion in the length of the bilayer appeared in figure 6 (b).

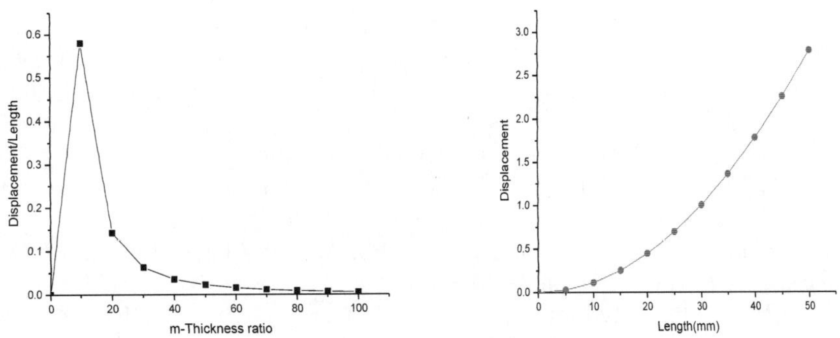

(a) (b)

Figure 6. Displacement of the actuator due to (a) Ratio of the displacement to length (b) Change in length.

In additionally investigated the some more prevailing parameter, for example, thermal expansion and temperature change with respect to the displacement of the bilayer actuator. The CTE of the two material decided for estimation. By expecting CTE of the two-layer difference to vary from zero to $10x10^{-4}/\ ^{0}C$. As per the theory, expository outcome gets that bigger the distinction of CTE could be prompt higher deflection is appeared in figure 7(a). At long last, a difference in temperature is additionally outlining the real part in bilayer actuator. Subsequently, figure 7(b) demonstrates the straight connection with a displacement of the actuator concerning the change in temperature increment. In this way, a thermal impact additionally affects the free end displacement of the bilayer actuator

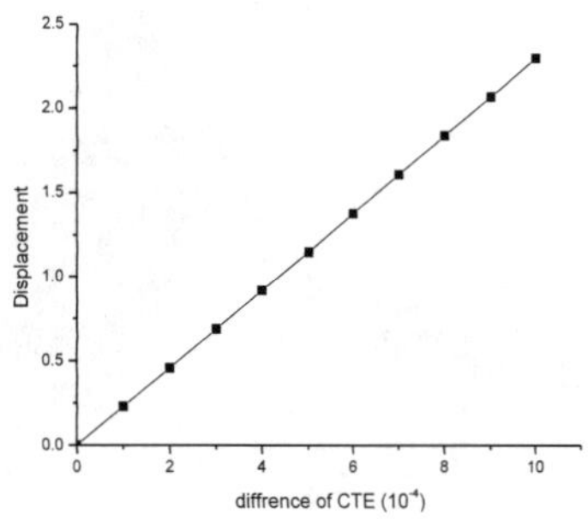

(a)

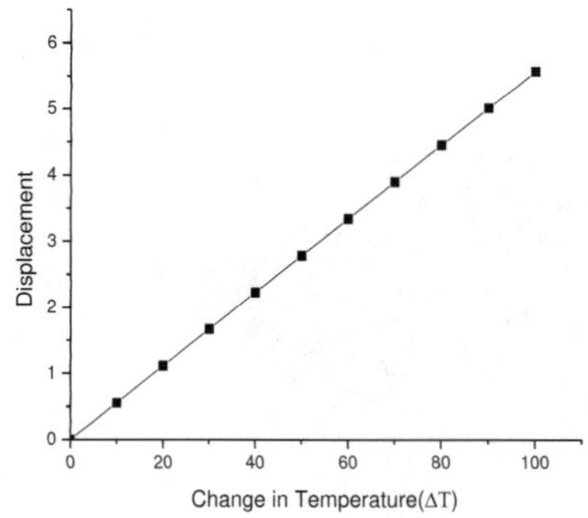

(b)

Figure 7. Actuation influence by (a) Difference in coefficient of thermal expansion (CTE) (b) Change in temperature.

5. Conclusion

In outline, the analytical results of bilayer actuator performance based on carbon black (CB) and PDMS composite is implemented effectively. The numerical outcome demonstrates that each layer thickness has a critical impact on the actuator deformation. Subsequently, the thickness of pristine PDMS and CB+ layer ought to be neither too substantial, nor too little. Likewise, bigger CTE difference and temperature change of the bilayer will predict more noteworthy deflection of the actuator. It found that deflection of bilayer actuator not just relies upon thickness likewise different parameters, for example, length, CTE, and temperature of the polymer composite material. This investigation is a reference for the design and fabrication of CB-polymer composite and furthermore a realization of experimental work because of thermal impact.

6. Reference

[1] Lim, H., Park, T., Na, J., Park, C., Kim, B. and Kim, E., 2017. NPG Asia Materials, 9(7), p.e399.
[2] Li, C., Liu, Y., Huang, X. and Jiang, H., 2012. Advanced Functional Materials, 22(24), pp.5166-5174..
[3] AbuZaiter, A. and Ali, M.S.M., 2014, 5th International Conference on (pp. 390-393). IEEE.
[4] Krulevitch, P., Lee, A.P., Ramsey, P.B., Trevino, J.C., Hamilton, J. and Northrup, M.A., 1996. Journal of Microelectromechanical Systems, 5(4), pp.270-282.
[5] Wang, H.S., Cho, J., Song, D.S., Jang, J.H., Jho, J.Y. and Park, J.H., 2017. ACS applied materials & interfaces, 9(26), pp.21998-22005.
[6] Duong, H.M., Papavassiliou, D.V., Mullen, K.J. and Maruyama, S., 2008. Nanotechnology, 19(6), p.065702.
[7] Hu, Y., Lan, T., Wu, G., Zhu, Z. and Chen, W., 2014. Nanoscale, 6(21), pp.12703-12709.
[8] Chen, L., Liu, C., Liu, K., Meng, C., Hu, C., Wang, J. and Fan, S., 2011. ACS nano, 5(3), pp.1588-1593.
[9] Liu, H., Niu, D., Jiang, W., Zhao, T., Lei, B., Yin, L., Shi, Y., Chen, B. and Lu, B., 2016. Sensors and Actuators A: Physical, 239, pp.45-53.
[10] Raturi, P., Kumar, A. and Singh, J.P., 2017. Smart Materials and Structures, 26(9), p.095030.
[11] Zeng, Z., Jin, H., Zhang, L., Zhang, H., Chen, Z., Gao, F. and Zhang, Z., 2015. Carbon, 84, pp.327-334.
[12] Jiang, W., Niu, D., Liu, H., Wang, C., Zhao, T., Yin, L., Shi, Y., Chen, B., Ding, Y. and Lu, B., 2014. Advanced Functional Materials, 24(48), pp.7598-7604.

[13] Wan, F., Luo, F., Wang, H., Huang, Z., Zhou, W. and Zhu, D.,2014. Ceramics International, 40(10), pp.15849-15857.

[14] Chen, H., Botef, I., Zheng, H., Maaza, M., Rao, V., and Srinivasu, V. V, 2011. International Journal of Nanotechnology, 8(6-7), 437-445.

[15] Pimentel-Domínguez, R., Velázquez-Benítez, A.M., Vélez-Cordero, J.R., Hautefeuille, M., Sánchez-Arévalo, F. and Hernández-Cordero, J., 2016. Polymers, 8(4), p.84.

[16] Timoshenko, S., 1925. JOSA, 11(3), pp.233-255.

IOP Conf. Series: Materials Science and Engineering **522** (2019) 012008 doi:10.1088/1757-899X/522/1/012008

Analytical and numerical modeling of a clamped-clamped micro beam under electric actuation

Aalim M Mustafa [1,*] **, Ahmed K Mahgoub** [2]

[1] Mechanical Engineering Department, Prince Mohammad bin Fahd University, Dhahran, Saudi Arabia
[2] Engineering Technology Program, PSEP, King Fahd University of Petroleum and Minerals, Dhahran, Saudi Arabia

[*] corresponding author's e-mail address: aalim.motasim@hotmail.com

Abstract. This paper studies the deflection of a clamped-clamped micro beam actuator subjected to an electric voltage. Two approaches were applied to model the micro beam under study: the first one is analytical and the second is numerical. A solution to the analytical model was developed to investigate the mechanical behaviour of the micro beam-based MEMS device under electric actuation and simulations was done using Wolfram Mathematica® software. On the other hand, a finite element model (FEM) simulation was performed using COMSOL Multiphysics® for data validation for the micro beam model, and the results showed very good agreement. The effect of changing beam length, thickness and height from the ground on both the maximum beam displacement and pull-in voltage was also analyzed.

1. Introduction

Micro-electromechanical systems (MEMS) are those systems or devices that contains very small mechanical and electrical components in micro scale. These devices are used to sense, control, and actuate to generate effects on the micro scale. In general, actuators are devices that convert an electrical signal into a mechanical force to manipulate other mechanical devices to perform some desired functions.

The dynamics of electrically actuated micro beams having clamped–clamped boundary conditions have been studied by many researchers in literature. Many researchers developed models using different techniques to investigate the dynamic behaviour and response of micro beams under different conditions. Younis et al [1] proposed a reduced-order model to study the behaviour of micro electro mechanical beams under electric actuation. The model was based on Galerkin method to discretize the equations of motion and convert it to finite number of ordinary differential equations in time domain. Ghayesh et al [2] investigated the behaviour of a micro electro mechanical systems using the theory of modified couple stress and the technique of pseudo-arclength continuation. They examined the dynamic response of the micro beam under primary and super-harmonic excitations. Mobki et al [3] studied the global stability of a micro beam hanged between two parallel conductive fixed plates. Different values of electric voltages that lead to significant variations in the behaviour of micro beam were obtained through bifurcation diagrams for different values of the gap between the micro beam and the ground. Younesian et al [4] examined the harmonic vibration of a clamped–clamped micro beam using the method of multiple scales. They scrutinized both the primary and the secondary resonances of the micro beam. Liu and Wang [5] studied the dynamic vibration of a clamped–clamped micro beam subjected to electrical voltage and affected by air damping. The displacement of the micro

Content from this work may be used under the terms of the Creative Commons Attribution 3.0 licence. Any further distribution of this work must maintain attribution to the author(s) and the title of the work, journal citation and DOI.
Published under licence by IOP Publishing Ltd

beam was studied under various actuating conditions and the effects of systems parameters on the center-point displacement of the micro beam were investigated. Bayat et al [6] used the Variational approach to analyze the vibration of an electrically actuated micro beam and provide analytical solutions for the significant vibrations of the micro beams.
The contribution of this paper is to develop a finite element model to find the deflection of a clamped-clamped micro beam using COMSOL Multiphysics software and compare the obtained numerical results with analytical solution available in literature. Moreover, the effect of the micro beam parameters on maximum beam's displacement and pull-in voltage will be investigated. This paper is organized as follows: Firstly, section 2 presents the governing equation of the clamped-clamped micro beam and the analytical solution of the model. Section 3 introduces the numerical validation using COMSOL Multiphysics software. Results will be presented and discussed in section 4, while section 5 reports a conclusion of the paper.

2. Mathematical model and solution

This study utilized the derived reduced order model presented by Younis et al [1] to study the static deflection of a clamped-clamped micro beam subjected to a constant DC voltage V loading as shown in figure (1).

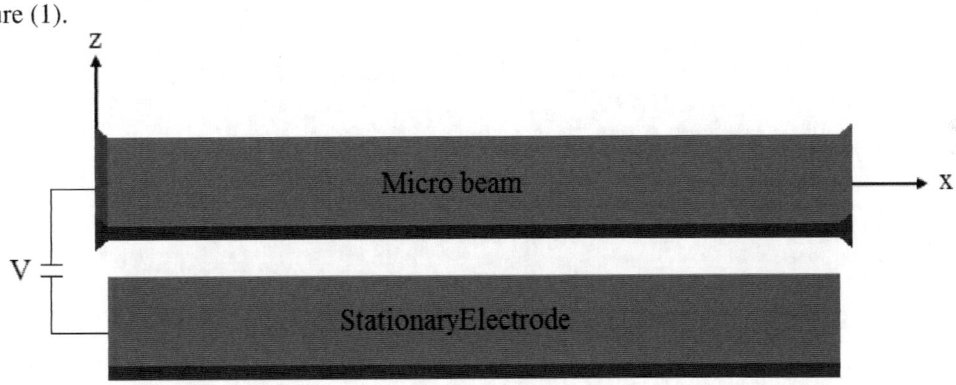

Figure 1. Electrically actuated clamped-clamped micro beam

The governing equation [1] of the static transverse displacement $w(x)$ is written as in equation (1)

$$EI\frac{d^4w}{dx^4} = \left[\frac{EA}{2L}\int_0^L (\frac{dw}{dx})^2\, dx\right]\frac{d^2w}{dx^2} + \frac{\epsilon\, b V^2}{2(d - w)^2} \tag{1}$$

Where x is the horizontal position across the beam length L, E is the modulus of elasticity, I is the second moment of inertia of the cross section, A is the cross section area of the micro beam, b is the micro beam's width, h is the micro beam thickness, ϵ is the dielectric constant of the gap medium and d is the gap width. For a clamped-clamped beam, the boundary conditions [1] are

$$w(0) = w(L) = 0$$
$$\frac{\partial w}{\partial x}(0) = \frac{\partial w}{\partial x}(L) = 0 \tag{2}$$

In order to solve these equations, one can introduce $w^* = \frac{w}{d}$, $x^* = \frac{x}{L}$ as a non-dimensional parameters to normalize the equation. Substituting the defined non-dimensional parameters into equation (1), dropping the star notations, and expanding the electric voltage term in a Taylor series up to fifth order, one can obtain

$$\frac{d^4w}{dx^4} = \left[\alpha_1 \int_0^L (\frac{dw}{dx})^2\, dx\right]\frac{d^2w}{dx^2} + \alpha_2 V^2(1 + 2w + 3w^2 + 4w^3 + 5w^4 + 6w^5) \tag{3}$$

Where the parameters α_1 and α_2 are given by $\alpha_1 = 6\left(\frac{d}{h}\right)$, $\alpha_2 = \frac{6\,\epsilon\, l^4}{E\, h^3 d^3}$. One can solve for the transverse displacement $w(x)$ of the micro beam by discretizing equation (3), using the first mode shape $\emptyset_1(x)$ of a clamped-clamped micro beam [7] as

$$\emptyset_1(x) = Cosh(4.73\,x) - Cos(4.73\,x) + 0.9825[Sin(4.73\,x) - Sinh(4.73\,x)] \tag{4}$$

$$w(x) = K\,\emptyset_1(x) \tag{5}$$

3. Model Validation with COMSOL

The finite element analysis software, COMSOL Multiphysics 4.4 [8] was used to develop a model of the micro beam to study it's deflection. Figure (2-a) displays the view of the micro beam model in COMSOL Multiphysics, while figure (2-b) shows the deflection along the length of the micro beam after running the simulation. When comparing the analytical results and the numerical results obtained from COMSOL Multiphysics, there is a very good agreement between the two models as shown in figure (3).

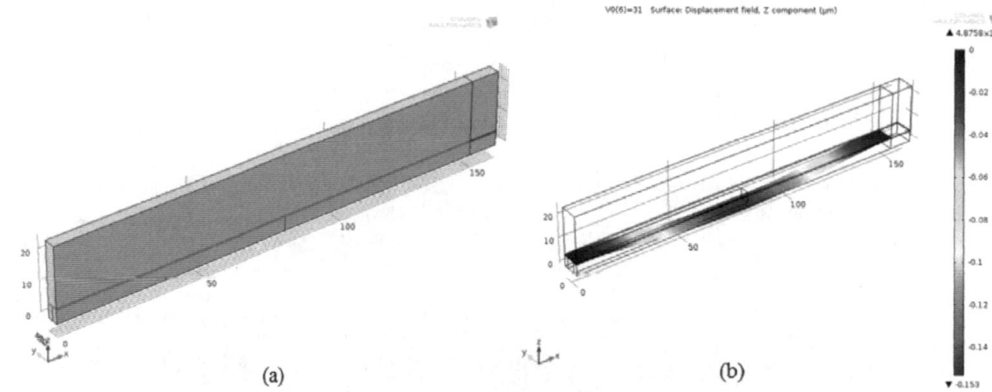

Figure 2. (a) micro beam model in COMSOL Multiphysics (b) displacement of the micro beam

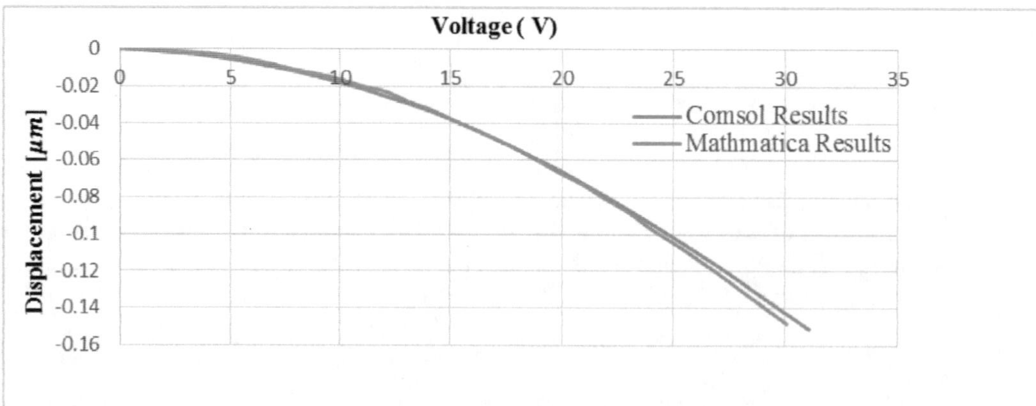

Figure 3. micro beam's maximum displacement under different voltages, Results Validation

4. Results and discussion

Consider the displacement of a micro beam under electric voltage. The micro beam under study is assumed to have the properties shown in table 1.

Table 1. Dimensions and material properties

Parameter	Value
d	$4.5\ \mu m$
b	$8\ \mu m$
h	$0.5\ \mu m$
L	$150\ \mu m$
ϵ	$8.854 * 10^{-12}$
E	$169 * 10^{9}$ Pa
ρ	$2332\ kg/m^3$
I	$\frac{bh^3}{12}$

4.1. Maximum displacement of the micro beam

To investigate the effect of the micro beam's parameters (length, thickness, height from the ground) on maximum beam displacement under different values of voltages, a parametric study is done by changing the parameter understudy and fixing all other parameters values.

Figure (4) shows the effect of voltage applied to the micro beam on maximum beam's deflection for different values of beam's lengths. It can be seen that increasing the applied voltage produces a significant increase in the maximum beam's displacement. In case of small voltages, i.e., up to 10 volts, changing beam's length affects the maximum beam's displacement very slightly. In details, when applying voltage of 10 V, changing beam's length from 80 μm to 150 μm produces a maximum displacement less than 0.02 μm for all three different lengths shown in figure (4), while at voltage of 30 V , changing beam's length from 80 μm to 150 μm increases the maximum displacement from 0.01 μm to 0.15 μm.

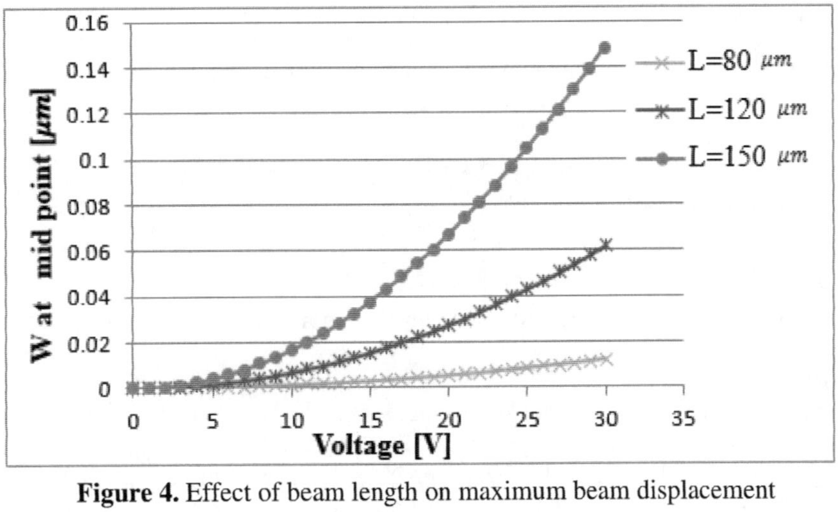

Figure 4. Effect of beam length on maximum beam displacement

Figure (5) demonstrates the effect of the micro beam thickness on it's maximum displacement for different values of electric voltage, V. It can be seen that increasing the thickness of the micro beam reduces the maximum displacement, especially when higher voltages is applied. For a voltage of 30 V, a reduction in thickness from *1 μm* to *0.5 μm*, increases the maximum displacement from *0.002 μm* to *0.15 μm*. In case of small voltages, the increase in micro beam's thickness affects the maximum displacement slightly.

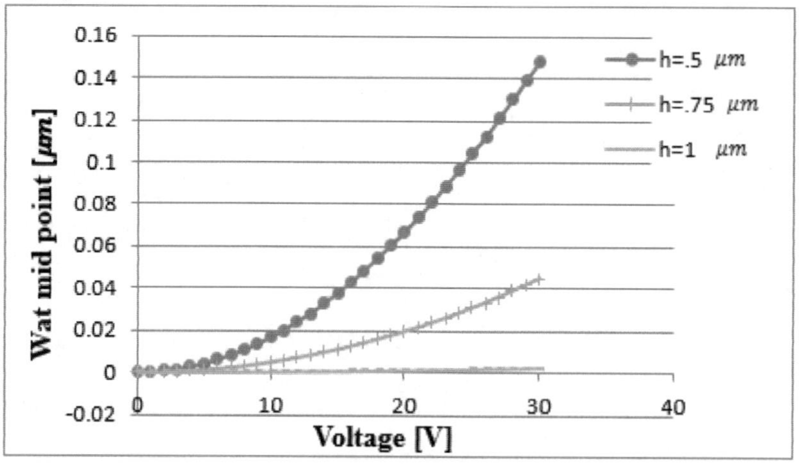

Figure 5. Effect of beam thickness on maximum beam displacement

Figure (6) shows the effect of the micro beam's height from the ground on it's maximum displacement for various values of electric voltage, V. It can be seen that increasing the height from the ground reduces the maximum displacement of the micro beam, especially when higher voltages is applied. For a voltage of 30 V, a reduction in thickness from *5 μm* to *2 μm*, increases the maximum displacement from *0.1 μm* to *0.7 μm*. In case of small voltages, the increase in micro beam's thickness affects the maximum displacement slightly.

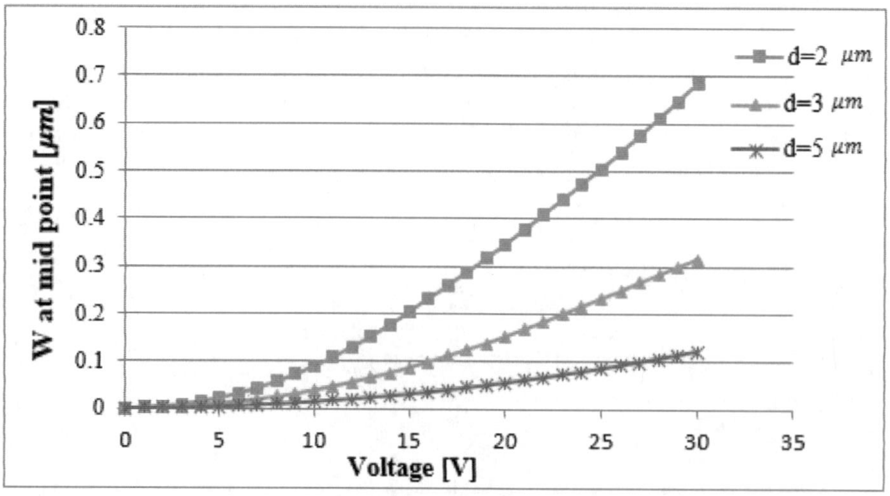

Figure 6. Effect of beam height form the ground on maximum beam displacement

4.2. Pull-in voltage

The term Pull-in voltage has been used widely in literature. The phenomenon of pull-in occurs when the deflection of the micro beam increases to the limit that the micro beam contacts the bottom electrode [9]. There are many studies published to investigate the pull-in phenomenon and its relation with parameters of MEMS devices. Chao et al [9] developed a correlation that predict the pull-in voltage of a clamped–clamped micro-beam using analytical solution of the partial differential equations of the dynamic model by plotting bifurcation and phase portrait diagrams. Pamidighantam et al [10] derived an expression to determine the pull-in voltage of clamped–clamped micro beams based on a lumped spring-mass system model. Farokhi et al [11] inspected the pull-in phenomenon occurrence in a curved clamped–clamped micro beam under actuation of an electrical voltage and investigated the effect of different system's parameters on pull-in voltage. Chen and Meguid [12] investigated the vibration of a micro beam electrically actuated by various values of DC and AC voltages. The governing equations of motion were developed and solved using the multiple scales method, and the frequency response of the micro beam was obtained.

In This section, authors will investigate the occurrence of the pull-in phenomenon for different values of beam's parameters. Figure (7a) presents the pull-in voltage for different values of beam's lengths. The pull-in voltage decreases with increasing the beam length, i.e., for a micro beam length of 150 μm the pull-in phenomenon will occur at voltage of 75 V, while if length of the beam increased to 300 μm the pull-in phenomenon will occur at lower voltage of 20 V. Figure (7b) shows that pull-in phenomenon occurs at low voltages for micro beams with low thicknesses. The pull-in voltage will increase from 20 V to 90 V as a result of increasing the thickness from 0.1 μm to 0.6 μm. Figure (7c) shows that pull-in phenomenon occurs at low voltages for micro beams close to the ground. The pull-in voltage will increase from 5 V to 90 V as a result of increasing the gap between the micro beam and its ground from 1 μm to 5 μm.

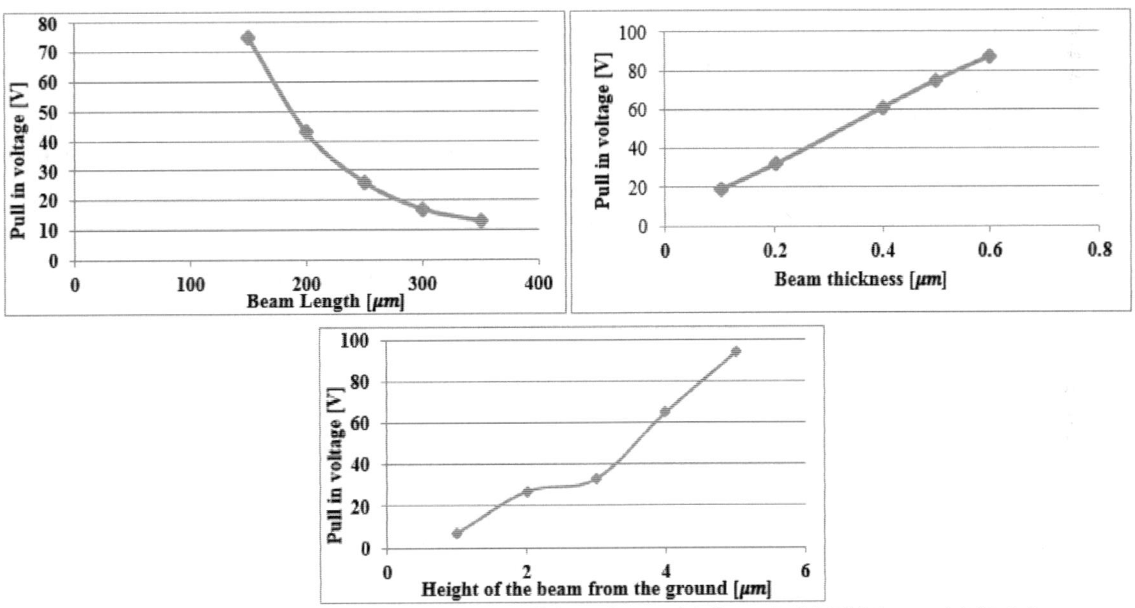

Figure 7. Effect of beam's parameters on pull-in voltage: (a) Length (b) Thickness (c) Height

5. Conclusion

The clamped-clamped micro beam actuator was modeled analytically and then a FEM simulation was performed for results validation and the obtained results showed a very good agreement. The effect of changing beam's parameters on its maximum displacement was analyzed by changing one parameters and keep other parameters at nominal values. Maximum beam's displacement increases with increasing beam's length while increasing beam's thickness and height from the ground reduce the maximum beam's displacement for the same applied voltage. Pull-in phenomenon was observed and the effect of beams parameters on the pull-in voltages was investigated.

Acknowledgments

The authors would like to acknowledge the support provided by Prince Mohammad bin Fahd University and King Fahd University of Petroleum and Minerals. The research was financially supported by Prince Mohammad bin Fahd University.

References

[1] Younis Mohammad I, Eihab M Abdel-Rahman and Ali Nayfeh. "A reduced-order model for electrically actuated microbeam-based MEMS." *Journal of Microelectromechanical systems* 12.5 (2003): 672-680.
[2] Ghayesh Mergen H, Hamed Farokhi and Marco Amabili. "Nonlinear behaviour of electrically actuated MEMS resonators." *International Journal of Engineering Science* 71 (2013): 137-155.
[3] Mobki Hamed, Ghader Rezazadeh, Morteza Sadeghi, Farid Vakili-Tahami and Mir-Masoud Seyyed-Fakhrabadi. "A comprehensive study of stability in an electro-statically actuated micro-beam." *International Journal of Non-Linear Mechanics* 48 (2013): 78-85.
[4] Younesian D, M Sadri and E Esmailzadeh. "Primary and secondary resonance analyses of clamped–clamped micro-beams." *Nonlinear dynamics* 76.4 (2014): 1867-1884.
[5] Liu Chin-Chia and Cheng-Chi Wang. "Numerical investigation into nonlinear dynamic behavior of electrically-actuated clamped–clamped micro-beam with squeeze-film damping effect." *Applied Mathematical Modelling* 38.13 (2014): 3269-3280.
[6] Bayat Mahdi, Mahmoud Bayat and Iman Pakar. "Nonlinear vibration of an electrostatically actuated microbeam." *Latin American Journal of Solids and Structures* 11.3 (2014): 534-544.
[7] Younis Mohammad I . "MEMS Linear and Nonlinear Statics and Dynamics" *Springer Science & Business Media* vol. 20. 2011.
[8] COMSOL Multiphysics User´s Guide.
[9] Chao Paul CP, C W Chiu and Tsu-Hsien Liu. "DC dynamic pull-in predictions for a generalized clamped–clamped micro-beam based on a continuous model and bifurcation analysis." *Journal of Micromechanics and Microengineering* 18.11 (2008): 115008.
[10] Pamidighantam Sayanu, Robert Puers, Kris Baert and Harrie AC Tilmans. "Pull-in voltage analysis of electrostatically actuated beam structures with fixed–fixed and fixed–free end conditions." *Journal of Micromechanics and Microengineering* 12.4 (2002): 458.
[11] Farokhi Hamed, Mergen H. Ghayesh and Shahid Hussain. "Pull-in characteristics of electrically actuated MEMS arches." *Mechanism and Machine Theory* 98 (2016): 133-150.
[12] Chen, X., and S. A. Meguid. "Nonlinear vibration analysis of a microbeam subject to electrostatic force." *Acta Mechanica* 228.4 (2017): 1343-1361.

Cement Substitutions in UHPC and their Influence on Principal Mechanical-Physical Properties

K Šeps[1], I Broukalová[1] and R Chylík[1]

[1] Faculty of Civil Engineering, CTU in Prague, Thákurova 7, 166 29 Prague, Czech Republic

karel.seps@fsv.cvut.cz

Abstract. The paper deals with supplementary cementitious materials (SCM) in ultra-high performance concrete and their influence on basic material characteristics. The supplementary materials were silica fume, fly-ash and metakaolin with dosage 10%, 20% and 30% of binder. The UHPC in concern is exceptional in use of aggregate with grains up to 16 mm in the mixture. The properties of hardened concrete were measured on cubes with size 100 mm and prism specimens 100x100x400 mm in age 28 days. The density, compressive strength and flexural strength of concrete with SCM were compared with reference concrete without supplementary materials.

1. Introduction

UHPC is a modern material that is increasingly used for construction. The predecessor of UHPC was high strength concrete with compressive strength 50 – 90 MPa largely used in concrete structures [1]. High strength was reached by application of additives as silica fume, fluid fly ash and ground blast furnace slag. Further enhancement of strength characteristic was reached by application of quartz and basalt aggregate [1, 2]. In 80ties of the last century the investigations of UHPC started. The research was initiated by increasing demands of special structures which required higher strength and environmental resistance of the material [3]. UHPC is very dense material with minimal pores and cavities; Its density is high ranging from 2500 to 3000 kg/m^3. Common compressive strength is 150 – 200 MPa. Tensile strength about 10 MPa is provided by low water/cement ratio and steel fibres dispersed in the cement matrix [4 - 6]. UHPC reaches high strengths thanks to high superplasticiser content and low w/c ratio and, the most important component affecting strength is cement. Cement dosage in UHPC is 800 – 1000 kg/m^3. Production of one kilogram of cement constitutes of 0.9 kg of CO_2, what represents ca 3.24 billions ton a year worldwide [7] and it corresponds with 7% of total CO_2 production. Application of substitution binders would reduce cement production; hence, the research meets both environmental and economic aspects.

Mostly, the UHPC is produced without coarse aggregate [8, 9]. The presented investigations focus in composite with aggregate grain size up to 16 mm, which distinguish the composite from other UHPC composites.

2. Materials and methods

In the experimental program, mixtures with supplementary cementitious materials were investigated. The indication of particular composites is following: metakaolin (MET), silica fume – microsilica

Content from this work may be used under the terms of the Creative Commons Attribution 3.0 licence. Any further distribution of this work must maintain attribution to the author(s) and the title of the work, journal citation and DOI.
Published under licence by IOP Publishing Ltd

(MIC), fly ash (FA) and reference mixture without and substitution of Portland cement. Composition of particular mixes is in the table 1. The substitutions are 10%, 20% a 30% of the cement weight. Water/binder ration is constant 0.26.

Table 1. Composition of the mixtures [kg/m³].

Compound	Ref	Met			Mic			FA		
		10 %	20 %	30 %	10 %	20 %	30 %	10 %	20 %	30 %
Cement	800	720	640	560	720	640	560	720	640	560
Water	210	210	210	210	231	252	273	197	184	172
Plasticizer	25.0	30.0	30.0.	30.0	33.0	33.0	33.0	34.0	32.0	30.0
Additive	-	80	160	240	80	160	240	80	160	240
w/b	0.26	0.26	0.26	0.26	0.26	0.26	0.26	0.26	0.26	0.26
Aggregate 0/4	730	730	730	730	730	730	730	730	730	730
Aggregate 4/8	390	390	390	390	390	390	390	390	390	390
Aggregate 8/16	320	320	320	320	320	320	320	320	320	320

For all mixes, specimens for measuring of compressive strength (cubes) and flexural strength (prisms) were elaborated. After demoulding the specimens were cured in air for 28 days. Than the specimens were measured and weighed and the density of concrete was calculated.
The compressive strength was determined according to ČSN EN 12390-3 [10] on four cubes with edge length 100 mm. Flexural tests were performed with prisms of dimensions 100x100x400 mm according to ČSN EN 12390-5 [11] in four-point bending tests without notch.

3. Results

Average values of densities are in the figure 1, compressive strength in a figure 2 and flexural strength in a figure 3.

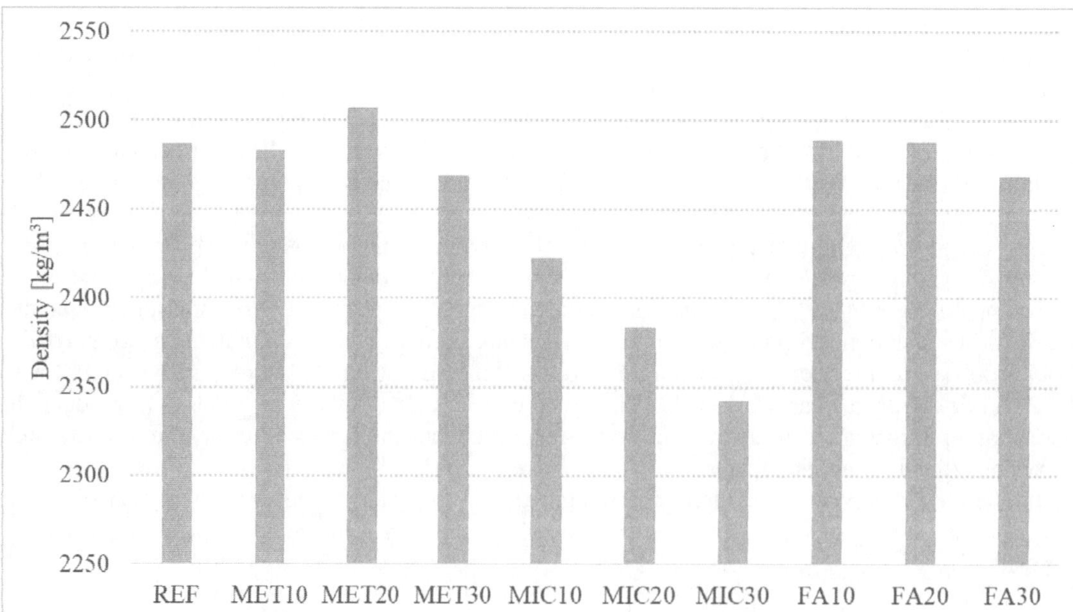

Figure 1. Average values of specimens´ densities.

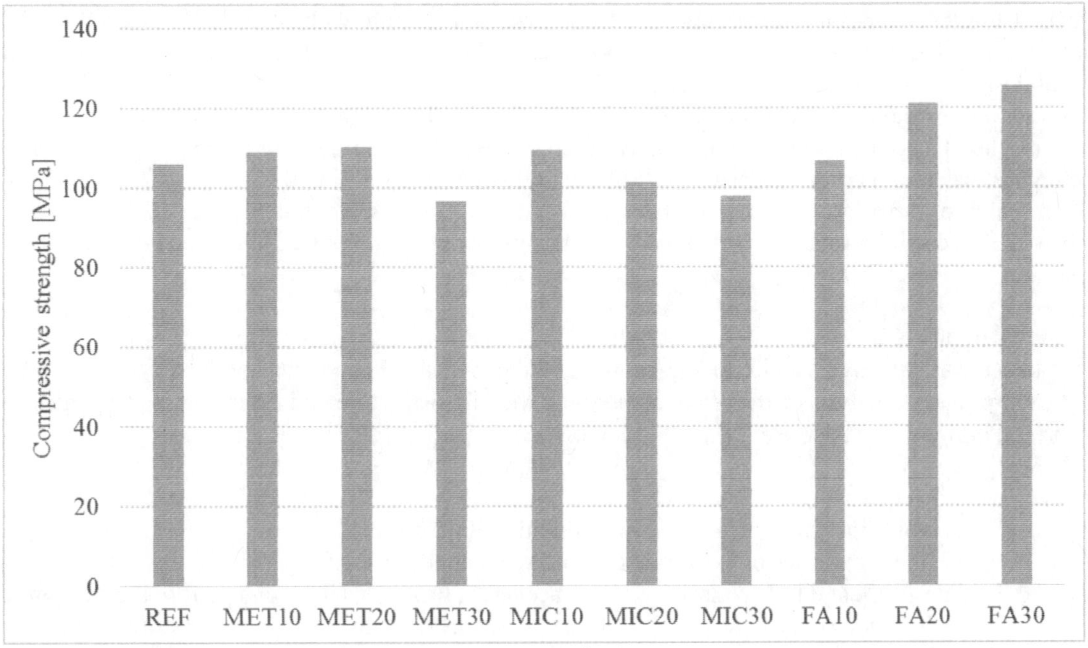

Figure 2. Average values of compressive strength measured on 100 mm cubes.

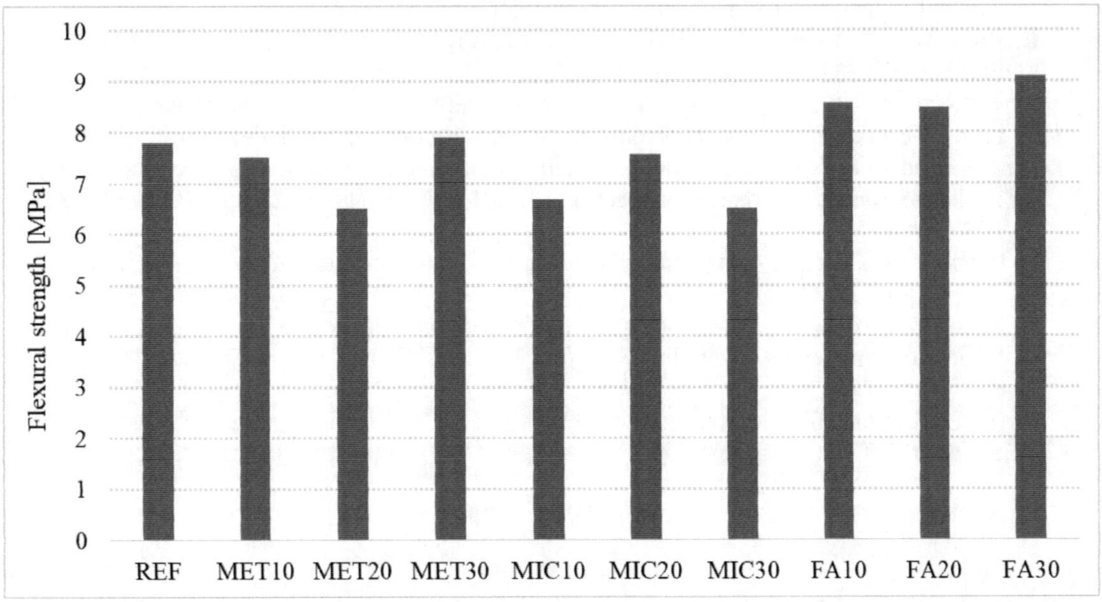

Figure 3. Average values of flexural strength measured in four-point bending on prism specimens 100x100x400 mm.

4. Conclusions

The results show that with increasing amount of supplementary materials, the concrete density lowers, which corresponds with the lower density of supplementary materials.
Compressive strength is higher for almost all mixes compared to reference mixture with exception of mixtures with 20% and 30% of silica fume and 30% of metakaolin.

Flexural strength from four-point bending test was lower for mixture with silica fume, metakaolin did not exhibit higher values with exception for 30% substitution. The fly ash increased flexural strength for all dosages of supplementary materials.
From achieved results a conclusion can be driven that for all investigated parameters, the best results were reached for fly-ash substitution, namely 20% or 30% substitution, where the composite provides the highest values of compressive strength and flexural strength.
The research program proved advisability of the Portland cement substitutions. Equal or even higher values are reached for reference mixtures, what brings saving of a considerable amount of Portland cement.

Acknowledgments
This paper was prepared thanks to the support of the Science Foundation of the Czech Republic (GAČR), project "Analysis of the relations between the microstructure and macroscopic properties of ultra-high performance concretes" (no. 17-19463S).

References
[1] S. Somayaji, Civil Engineering Materials, Prentice Hall, 2001.
[2] A.M. Neville, Properties of Concrete, fourth ed., Prentice Hall, 2005.
[3] E. Fehling, M. Schmidt, J. Walraven, T. Leutbecher, S. Fröhlich, Ultra-high performance concrete UHPC: Fundamentals-Design-Examples, Wilhelm Ernst & Sohn, 10245 Berlin, Germany, 2014
[4] Y. Su, C. Wu, J. Li, Z.X. Li, W. Li, Development of novel ultra-high performance concrete: from material to structure, Constr. Build. Mater. 135 (2017) 517– 528.
[5] C. Shi, Z. Wu, D. Wang, J. Xiao, Z. Huang, Z. Fang, A review on ultra high performance concrete: Part I. Raw materials and mixture design, Constr. Build. Mater. 101 (2015) 741–751.
[6] A. Kohoutková, P. Bílý, J. Fládr, J. Vodička, Experimental verification of the behavior of protective steel liner of concrete containment, Key Engineering Materials 691, (2016) 3-13.
[7] A.M. Rashad, D.M. Sadek, H.A. Hassan, An investigation on blast-furnace slag as fine aggregate in alkali-activated slag mortars subjected to elevated temperatures, J. Clean. Prod. 112 (2016) 1086–1096.
[8] P. Richard, M. Cherezy, Composition of reactive powder concrete, Cem. Concr. Res. 25 (7) (1995) 1501–1511.
[9] V. Matte, M. Moranville, Durability of reactive powder composites: influence of silica fume on the leaching properties of very low water/binder pastes, Cement Concr. Compos. 21 (1999) 1–9.
[10] ČSN EN 12390-3 Testing hardened concrete – Part 3: Compressive strength of test specimens
[11] ČSN EN 12390-5 Testing hardened concrete – Part 5: Flexural strength of test specimens

IOP Conf. Series: Materials Science and Engineering **522** (2019) 012010 doi:10.1088/1757-899X/522/1/012010

Analysis of the influence of supplementary cementitious materials used in UHPC on modulus of elasticity

J Fládr[1], P Bílý[1] and T Trtík[1]

[1] Faculty of Civil Engineering, CTU in Prague, Thákurova 7, 166 29 Prague, Czech Republic

josef.fladr@fsv.cvut.cz

Abstract. The paper presents the results of static and dynamic modulus of elasticity measurements that were conducted on high-performance concretes with partial replacement of cement by supplementary cementitious materials (SCM). Metakaolin, microsilica and fly ash were used as SCM, the replacement levels were 10 %, 20 % and 30 % of cement weight. The tests were conducted on cylindrical samples 100 mm in diameter and 200 mm in height. Dynamic modulus was measured at the age of 1, 3, 7, 14, 21 and 28 days with the use of non-destructive ultrasound pulse method. Static modulus was determined by destructive compression test. The highest values were reached for fly ash at 20 % and 30 % replacements. Samples containing metakaolin had always lower modulus than the reference concrete. The coefficients for recalculation of dynamic modulus to static modulus were derived as 0.95 for reference concrete, 0.80 for metakaolin concrete, 0.85 for microsilica concrete and 0.90 for fly ash concrete.

1. Introduction

Cement is the most expensive compound of concrete and at the same time its production yields the biggest amounts of greenhouse gases [1]. The research of possibilities of alternative cement substitutions can reduce the costs of concrete production and improve the environment. The important benefit is the exploitation of waste materials as cement replacement, because it also reduces the landfilling. The examples of such materials are microsilica or fly ash. Both these additives can be classified as latent hydraulic ones and therefore they are suitable as supplementary cementitious materials (SCM) [2].

When substituting cement by SCM, it is important to check the mechanical properties of concrete as both their values and time development can be changed. This fact is widely known and supervised in the industrial production. Unfortunately, only the compressive strength is usually observed. In some cases the durability is also considered. Modulus of elasticity is often neglected even in case of concrete without additives. At the same time, the value of modulus of elasticity is very important for the design of structures to serviceability limit states. Moreover, the rate of increase of modulus of elasticity significantly influences the speed of construction process of in-situ cast structures.

2. Experimental methods

In the experimental program, three types of SCM were used: microsilica (labelled as Mic), fly ash (FA) and metakaolin (Met). All these SCM were used as partial replacement of Portland cement by 10 %, 20 % and 30 % of cement weight. The composition of the mixtures is given in table 1. For the reference, the mixture labelled as Ref with no SCM was used.

Content from this work may be used under the terms of the Creative Commons Attribution 3.0 licence. Any further distribution of this work must maintain attribution to the author(s) and the title of the work, journal citation and DOI.
Published under licence by IOP Publishing Ltd

IOP Conf. Series: Materials Science and Engineering **522** (2019) 012010 doi:10.1088/1757-899X/522/1/012010

Table 1. Composition of the mixtures [kg/m^3].

Compound	Ref	Met			Mic			FA		
		10 %	20 %	30 %	10 %	20 %	30 %	10 %	20 %	30 %
Cement	800	720	640	560	720	640	560	720	640	560
Water	210	210	210	210	231	252	273	197	184	172
Plasticizer	25.0	30.0	30.0	30.0	33.0	33.0	33.0	34.0	32.0	30.0
Additive	-	80	160	240	80	160	240	80	160	240
w/b	0.26	0.26	0.26	0.26	0.26	0.26	0.26	0.26	0.26	0.26
Aggregate 0/4	730	730	730	730	730	730	730	730	730	730
Aggregate 4/8	390	390	390	390	390	390	390	390	390	390
Aggregate 8/16	320	320	320	320	320	320	320	320	320	320

Three cylinders (100 mm diameter, 200 mm height) and three 100 mm cubes were produced from each mixture. The values given in the paper are arithmetic averages of three measurements in all cases. The specimens were unmoulded and dynamic modulus measurements were commenced one day after concreting. Further measurements took place at 3, 7, 14, 21 and 28 days. During this time, the samples were stored at laboratory conditions (25 °C, 30 % relative humidity). This is not in accordance with the guidelines for storage of specimens for strength tests [3] which require the relative humidity to be between 95 – 100 %. The authors wanted to avoid the fact that if the samples were stored in water, the pores would be filled by water which would significantly affect the dynamic modulus measurements.

2.1. Dynamic modulus of elasticity

The determination of dynamic modulus was carried out by non-destructive ultrasonic pulse method according to [4]. Pundit Lab (+) device by Proceq company was used together with 54 kHz probes and sonogel improving the contact of probes with the sample.

Before each measurement of velocity of passage of the pulse through the sample, the sample was weighted with 0.01 g precision. The current weight of the sample was used for calculation of dynamic modulus according to equation (1). The measured values were also important for humidity decrease rate control.

$$E_{dyn} = \frac{\rho \cdot \left(\frac{v_u}{1000} \right)^2}{k^2 \cdot 1000} \tag{1}$$

In the equation, ρ is bulk density [kg/m^3], v_u is the velocity of the ultrasound pulse wave [m/s] and k is the largeness coefficient [-] according to equation (2).

$$k = \sqrt{\frac{1-\nu}{(1+\nu)\cdot(1-2\cdot\nu)}} \tag{2}$$

The Poisson ratio (ν) was considered as 0.2. The final values of dynamic modulus of elasticity are listed in table 2.

2.2. Static modulus of elasticity

The values were determined by destructive compression tests. Before the main test, it was necessary to establish the value of compressive strength for all types of concrete. The cubic compressive strength, which is important for proper set up of loading cycles, was measured on 100 mm cubes and multiplied by 0.8 to obtain cylindrical compressive strength (in accordance with previous experience of the authors and literature recommendations [5]).

The cylinder bottoms were ground before the tests to secure planarity and collinearity of both bottoms. The specimens were equipped with two independent strain gauges and tested according to [6]. Average results are given in table 2.

3. Results

The time development of dynamic elastic modulus of all the mixtures is given in figures 1 – 3 for particular additives. Reference concrete is inserted for comparison in all the figures.

Table 2 summarizes the values of dynamic and static modulus at the age of 28 days. The last row of the table shows the ratio between static and dynamic modulus of elasticity.

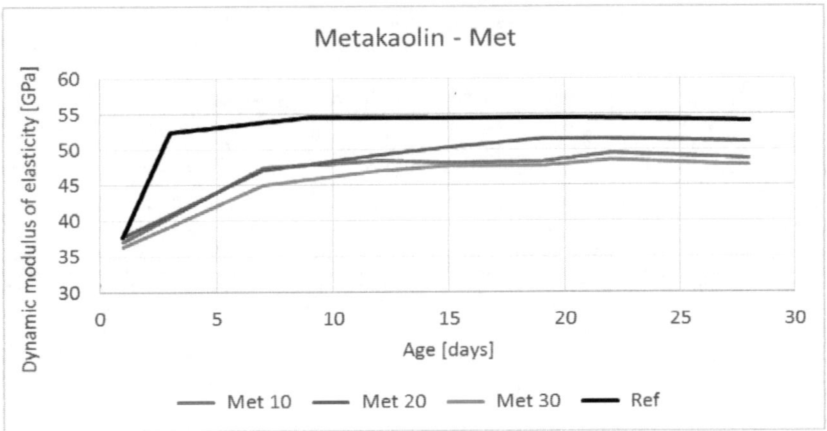

Figure 1. Time development of dynamic modulus of elasticity – mixtures containing metakaolin.

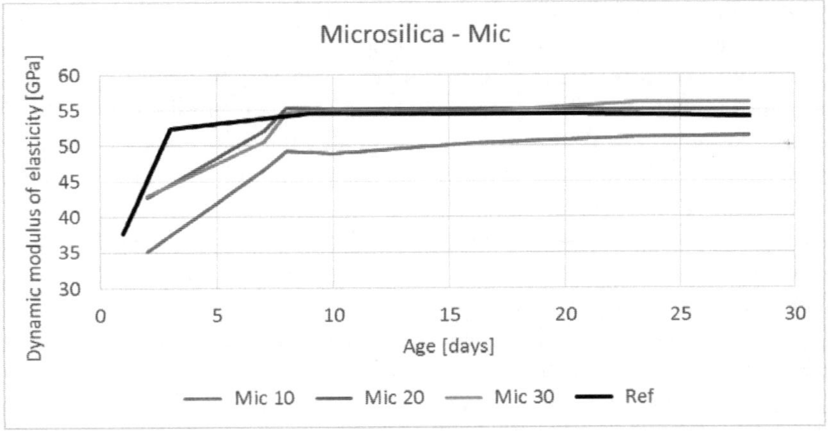

Figure 2. Time development of dynamic modulus of elasticity – mixtures containing microsilica.

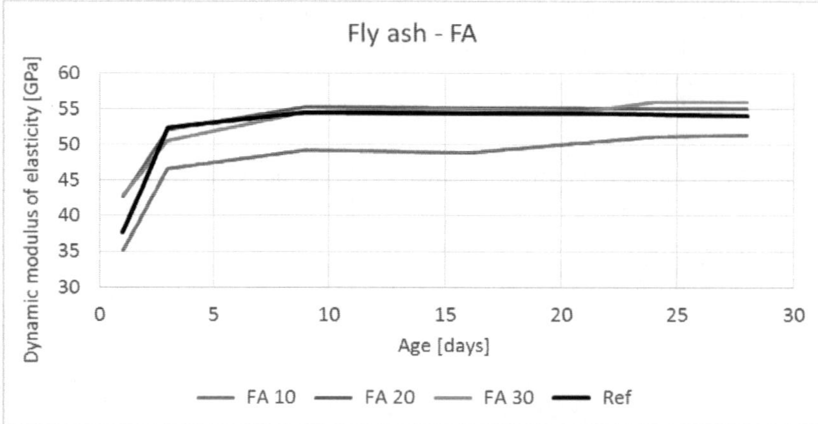

Figure 3. Time development of dynamic modulus of elasticity – mixtures containing fly ash.

Table 2. Values of dynamic and static modulus of elasticity at the age of 28 days [GPa].

Modulus	Ref	Met			Mic			FA		
		10 %	20 %	30 %	10 %	20 %	30 %	10 %	20 %	30 %
Dynamic (E_d)	54.09	51.34	55.02	55.95	51.34	55.02	55.95	48.73	51.15	47.85
Static (E_s)	51.30	42.16	46.76	48.10	46.20	49.51	49.78	38.98	41.94	39.71
E_s/E_d	0.95	0.82	0.85	0.86	0.90	0.90	0.89	0.80	0.82	0.83

4. Conclusions

From the results presented in the previous section, following conclusions regarding the changes of values of modulus of elasticity can be drawn:

- The development of dynamic modulus was significant in the first 10 days, later the value was practically constant for all the mixtures.
- Metakaolin reduced the value of dynamic modulus at all replacement levels (10 %, 20 %, 30 %). At the same time, the growth rate was smaller than in case of reference concrete.
- Microsilica at 20 % and 30 % replacement showed the same values of dynamic modulus from the age of 10 days until the end of the measurement. The samples with 10 % replacement had systematically lower modulus than the reference concrete.
- The rate of dynamic modulus growth of microsilica concrete was smaller than for reference concrete at all replacement levels.
- Fly ash at 20 % and 30 % replacement behaved in the same way as the reference mixture. One day value was even higher than for the reference concrete.
- Fly ash at 10 % replacement had lower elastic modulus than the reference concrete during whole course of the measurement campaign.
- The recalculation coefficients from dynamic to static elastic modulus at the age of 28 days were dependent on the type of additive only, not on the replacement level. For the reference concrete the value was 0.95, for metakaolin concrete it was determined as 0.80, for microsilica concrete as 0.85 and for fly ash concrete as 0.90.

Based on the aforementioned conclusions, it can be said that to reach the highest possible modulus of elasticity of concrete with SCM, the best solution is to replace 20 % or 30 % of cement weight by fly ash. An agreement between the dependence of dynamic modulus and static modulus on cement replacement level was observed.

Acknowledgments

This paper was prepared thanks to the support of the Science Foundation of the Czech Republic (GAČR), project "Analysis of the relations between the microstructure and macroscopic properties of ultra-high performance concretes" (no. 17-19463S).

References

[1] Kara M 2012 Environmental and economic advantages associated with the use of RDF in cement kilns Resources, *Conservation and Recycling* vol 68 p 21-28
[2] Chylík R and Šeps K 2018 Influence of cement replacement by admixture on mechanical properties of concrete *Proceedings of the 12th International fib PhD symposium in Civil Engineering Calorimetry* (Prague, Czech Republic: Czech Technical University) Currently in print
[3] ČSN EN 12390-2, Testing hardened concrete - Part 2: Making and curing specimens for strength tests.
[4] ČSN EN 12504-4, Testing concrete - Part 4: Determination of ultrasonic pulse velocity.
[5] Collepardi M 2010 *The New Concrete*, vol 2, Grafiche Tintoretto p 436 ISBN: 8890377720
[6] ČSN EN 12390-13, Testing hardened concrete - Part 13: Determination of secant modulus of elasticity in compression.

Effect of GGBS and Nano Silica on the Mechanical Properties of Ternary Concrete

Manank Shah[1], Niragi Dave[1], Kaushal Vora[1], Dhruv Shah[1]

[1] School of Technology, Pandit Deendayal Petroleum University, Gandhinagar, Gujarat- 382007

Corresponding author - e-mail: zizkova.n@fce.vutbr.cz

Abstract. This paper presents the effect of mechanical behavior of high strength concrete, prepared using Ground Granulated Blast Furnace Slag (GGBS) and Nano Silica (Average particle size-15nm). Ground Granulated Blast Furnace Slag (GGBS) was replaced up to 50% and Nano silica was replaced 1%, 2%, 3%, 4% and 5% by weight of cement. Compression Testing, Pull Out Testing, Split Tensile Testing and Flexural Testing were carried out to characterize the mechanical behavior of high strength concretes at 7, 28, 56 and 90 days and results obtained were compared with the corresponding values obtained for controlled concrete (100% OPC). The entire specimen were prepared, cured and tested as per the Indian standard code of practice. The results showed that compressive strength significantly improved at early age with the increasing content of Nano silica. Compressive, tensile and flexure strength of the concrete with GGBS and NS was found better than that of the controlled concrete. Utilization of this type of material into the public sphere through civil infrastructure can be done by its usage in Dams and High rise buildings where dense concrete is required.

1. Introduction

Nanotechnology has been regarded as an emerging technology, introducing new dimensions to science and technology with the possibility of manipulating atoms and molecules at the nanometer level ('nano' means one-billionth of a meter) [1]. Nano technology can be considered as most advanced field of science and technology and has great market potential and economic impacts [2]. The aim of this research work is to understand the behavior of materials at nano level and how to improve microstructure of the cementitious materials and to achieve cohesiveness in concrete structures. Concrete is most used man made material on this earth and worldwide production of concrete has exceeded over 2600 million tones [3]. It is responsible for about 5% of the global man made CO2 emission. For each tone of cement being produced, an average of 0.87 tons of CO2 is being emitted [4, 5]. Extensive research has been directed to reduce the effect of cement industry on green house gases [6].

GGBS, a by- product of Iron industry is considered as one of the most used SCM in the industry. The use of supplementary cementitious materials as partial replacement to Portland cement in concrete is a better step towards sustainable development because of their technological, economic, and environmental benefits

2. Materials and methods

Materials used in this research work are: OPC (53 grade) conforming to IS: 8112-1989 and Ground Granulated Blast furnace Slag (GGBS) whose chemical and mineralogical properties are shown in

Content from this work may be used under the terms of the Creative Commons Attribution 3.0 licence. Any further distribution of this work must maintain attribution to the author(s) and the title of the work, journal citation and DOI.
Published under licence by IOP Publishing Ltd

Table 1. Nano Silica (Hydrophilic Nano powder, Reinste Nano Ventures) of average particle diameter 10 nm has been substituted 1%, 2%, 3%, 4% and 5% by weight of cementitious material. In the present study, Super Plasticizer (SP) of Carboxylic Ether (CONFLOW SNS 2, Essence Construction) with specific gravity 1.21 was used. Table 2 demonstrates the mix design of M 60 grade of concrete used during the study.

The gravel used was crushed gravel with size less than 12.5 mm in accordance to ASTM standard grading curve. The fine aggregate used was river sand conforming to IS: 383 and passing from sieve size of 4.75 mm. The of river sand of specific gravity 2.67, Bulk Density 1585 Kg/m^3 and fineness modulus of 2.62 was used in the research work.

Compressive strength was measured as per I.S. 516-1959 using 200 T capacity universal testing machine using standard 150×150×150 mm cubes. Concrete cylinders of size 150 mm diameters and 300 mm height were casted. After the specified curing period was over, the concrete cylinders were subjected to split tensile test by using universal testing machine. Tests were carried out on triplicate specimens and the average split tensile strength values were recorded. Flexure strength was measured using standard 100×100×500 mm beam specimens, simply supported on an effective span of 400 mm and loaded at the third points after 28 days of curing. The test was carried out as per IS: 516-1959 (Reaffirmed 2004). The Bond strength of the concrete was measured using 16 mm TMT bar as per Indian standard (IS:2770-1) code of practice and results obtained for M60 grade concrete for different binders at 28, 56, 90 days of curing The pull out specimens were then tested in the hollow mechanical machine of 2000 KN capacity

Table 1. Chemical and mineralogical properties of OPC, GGBS and NS

Description	Water	OPC	GGBS	Nano Silica
Physical Characteristics				
Specific Gravity	1	3.15	3.14	2.4-2.6
Blaine's Fineness, cm^2/gm	-	2285	3250	10,00,000
Chemical Characteristics				
Calcium oxide, Cao, %	-	66.73	35.9	-
Silicon dioxide, SiO2, %	-	17.54	40.65	99.9
Aluminum oxide, Al2O3, %		9.82	17.07	-
Ferric oxide, Fe2O3, %	-	2.19	0.68	-
Manganese oxide, MnO, %	-	0.02	0.03	-
Magnesium oxide, MgO, %	-	1.24	3.75	-
Potassium oxide, K2O, %	-	0.48	0.56	-
Sodium oxide, Na2O, %	-	0.22	0.19	-
Loss of ignition, %	-	0.9	0.08	-

Table 2. The mix design concrete samples used in study

		Kg/m^3						
Mix No.	Mix	Cement	GGBS	Nano Silica	Coarse Aggregate	Fine Aggregate	Water	SP
1	100% OPC	594	0	0	755.5	807.5	208	8.07
2	50% OPC+50% GGBS	297	297	0	755.5	807.5	208	8.07
3	49% OPC+50% GGBS+1% NS	291.06	297	5.94	755.5	807.5	208	8.07
4	48% OPC+50% GGBS+2% NS	285.12	297	11.88	755.5	807.5	208	8.07
5	47% OPC+50% GGBS+3% NS	279.18	297	17.82	755.5	807.5	208	8.57
6	46% OPC+50% GGBS+4% NS	273.24	297	23.76	755.5	807.5	208	8.57
7	45% OPC+50%	267.3	297	29.7	755.5	807.5	208	8.98

	GGBS+5% NS								

3. Results and discussion:

3.1. Study of workability.
Table 3 indicates the values of slump in mm of the concrete mix used in this study. It can be noted from Table 3 that the range of the slump varies from 120-166 mm. During the test, it was noted that further addition of Nano Silica makes concrete sticky and also decreases its workability. It can be observed that mix 6 has highest slump value as compared to others.

Table 3. Workability of M60 grade of concrete

MIX COMBINATIONS	MIX NO.	SLUMP (MM)
100% Control	Mix1	120
50% OPC+50% GGBS	Mix2	127
49% OPC+50% GGBS+1% NS	Mix3	138
48% OPC+50% GGBS+2% NS	Mix4	146
47% OPC+50% GGBS+3% NS	Mix5	154
46% OPC+50% GGBS+4% NS	Mix6	166
45% OPC+50% GGBS+5% NS	Mix7	157

3.2. Compressive Strength Test.
The averaged compressive strength results of cube specimens of each batch of different concrete cubes were measured at 7, 28, 56 and 90 days of curing are presented in Fig.1. It can be observed that compressive strength of all mix combinations increases over time. Mix 6 and 7 have shown better performance among others at higher ages. In all mixes samples, when Nano silica content increased to an optimum value of 4% in the sample, compressive strength of the sample was significantly improved. This is due to the Pozzolanic reaction of Nano silica, its high surface area reactive with C3S and C2S and produces C-S-H condensed gel as a result of reaction with Ca(OH)2. Nano particle, having filling ability property, fills the porosity in C-S-H gel and makes a denser adhesive cement paste. This reduction is because the Nano particles have high surface area and are likely to stick together through a physical reaction, and form unstable balls. The sticky mix which reduces the workability of concrete and hampers mixing and compaction process. It leads to the reduction in the compressive strength. Thus 4% is the optimum percentage. This trend is observed at all ages and the strength for different percentages of Nano silica at different ages is higher than that of the reference.

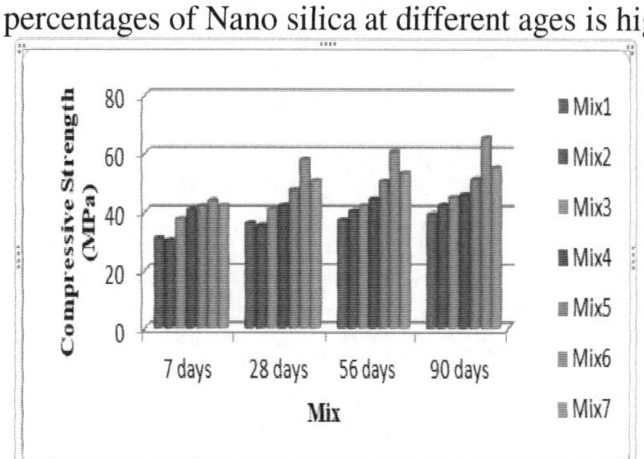

Figure 1 Compressive Strength of M60 grade of Concrete

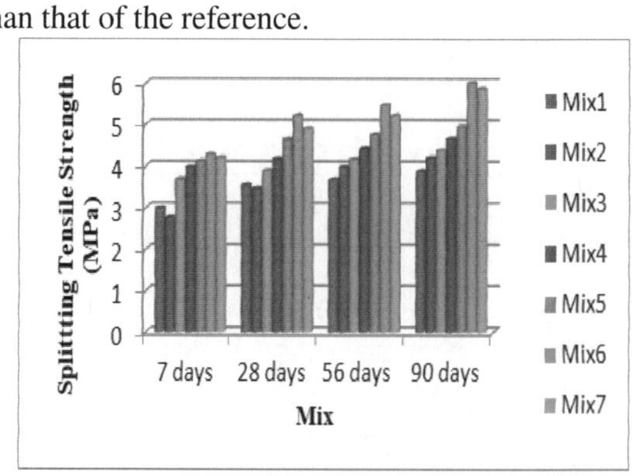

Figure 2 Tensile Strength of M60 grade of concrete

3.3. Tensile Strength Test.

It was also observed that the tensile strength of all mix combinations increases over time as evident in Fig. 2. The increase in the split tensile strength as compared control mix can be attributed to continuation pozzolanic reactions with time. Moreover, high surface area of the Nano silica decreases the air void and evidently cause increase in the Tensile strength. It can be observed from Fig.2 that for Mix 6, increase of 46.7%, 49.31%, 54.63% of tensile strength of OPC could be seen, which is maximum among all other mixes. 4% Nano Silica yields optimum Tensile Strength for all ages. Also, it was concluded that split tensile strength of concrete does not increase significantly with replacement of very high percentage of Nano Silica.

3.4. Flexural Strength Test.

Fig.3 summarizes the flexural strength of unitary, binary, and ternary concretes at 7, 28, 56 and 90 age days of curing. It can be observed from Fig.3 that for Mix 6, increase of 21.28%, 35.10%, 36.94%, 48.24% of flexural strength of OPC could be seen, which is maximum among all other mixes. It was observed from the test results that flexural strength increases till 4% replacement of Nano silica and thereafter there is no significant increase in the flexural strength. . Flexural strength assessments demonstrated that having filler and pozzolanic effects, Nano silica can improve the structural properties of the concrete.

It is vital that you do not add any headers, footers or page numbers to your paper; these will be added during the production process at IOP Publishing (this is why the Header and Footer margins are set to 0 cm in table 1).

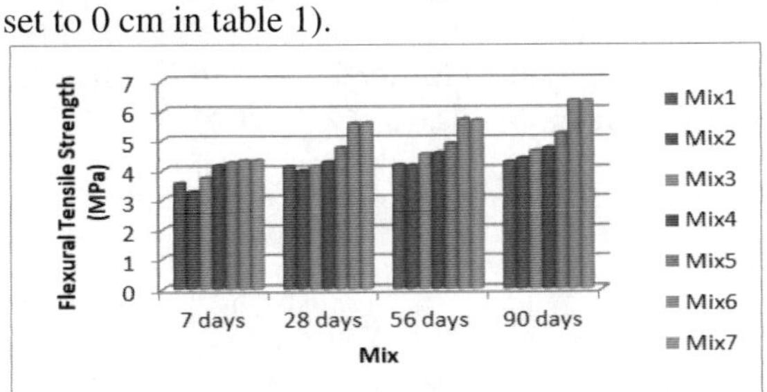

Figure 3 Flexural Strength of M60 grade of concrete.

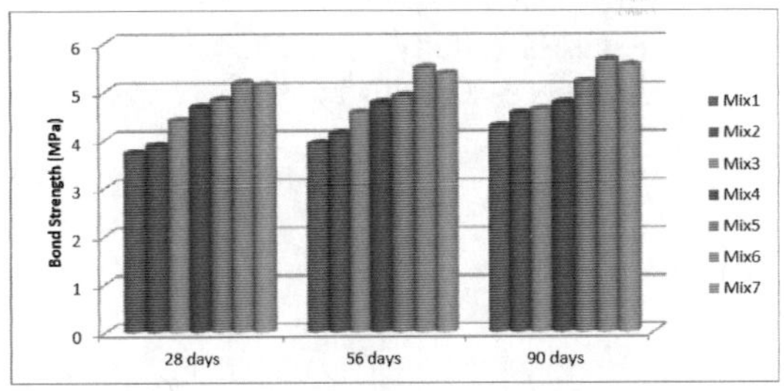

Figure 4 Bond Strength of M60 grade of concrete

3.5. Pull Out Strength (Bond Strength)

The pullout test specimens failed by the following three modes of failures, pullout failure (P), splitting failure of the tested specimen (S) and steel rupture failure (CS). The pullout failure mode occurred when the concrete cover provided adequate confinement, thus preventing a splitting failure of the test specimen. Splitting failure mode was the predominant type of failure of the tested specimen. It was characterized by splitting of the concrete specimen in a brittle mode of failure. Both transverse and longitudinal cracks were observed at failure. It can be observed from fig. 4 result of bond strength test that the mix which had maximum cohesiveness between the concrete and steel bars due to minimum air voids has maximum bond strength.

4. Conclusions

This study, experimentally evaluated the combined effects of GGBS and Nano silica on the mechanical properties of high strength concrete. The following conclusions can be drawn from this study:

Workability of the concrete mix is observed to increase till maximum of 4% replacement of Nano silica and thereafter workability was reduced. Reduction in workability can be attributed to high surface area of nano silica and formation of sticky concrete paste.

Bond Strength was found maximum for ternary mix (50% GGBS, 46% OPC and 4% Nano Silica). The increase in bond strength of the mix as compared to control can be attributed to increased cohesiveness as compared to control.

IOP Conf. Series: Materials Science and Engineering **522** (2019) 012011 doi:10.1088/1757-899X/522/1/012011

Mechanical properties such as compressive, flexural and tensile strength are initially increased by the increase of Nano silica content up to 4% and then decreased. The increase in strength is because of Nano silica pozzolanic reactions while its decrease is because of high surface area of Nano particles. When the content of Nano particles increases beyond a certain value (optimal percentage), they will stick to each other through a physical reaction leading unstable balls. An increase of 4% of Nano silica is considered as optimal value in this experimental study.

References

[1] Kumiko Miyazaki, Nazrul Islam "Nanotechnology systems of innovation—An analysis of industry and academia research." Technovation(Elsevier) 27 (2007) 661–675.

[2] A.M. Said, M.S. Zeidan, M.T. Bassuoni, Y. Tian " Properties of concrete incorporating nano-silica A.M. Said." Construction and Building Materials(Elsevier).

[3] USGS.org [Internet]. Reston, VA: United States Geological Survey; c2010 [cited 2011 May 9]. < http://minerals.usgs.gov/minerals/pubs/commodity/cement/mcs-2010-cemen.pdf >.

[4] J.S. Damtoft, J. Lukasik, D. Herfort, E.M. Gartner "Sustainable development and climate change initiatives." Cement and Concrete Research(Elsevier) 38 (2008) 115–127.

[5] D.J. Barker, S.A. Turnera , P.A. Napier-Moore , M. Clark and J.E. Davison."CO2 Capture in the Cement Industry." Energy Procedia(Elsevier) 1 (2009) 87-94.

[6] Niragi Dave , Anil Kumar Misra , Amit Srivastava , Anil Kumar Sharma , Surendra Kumar Kaushik "Study on quaternary concrete micro-structure, strength, durability considering the influence of multi-factors."Construction and Building Materials(Elsevier) 139 (2017) 447-457.

IOP Conf. Series: Materials Science and Engineering **522** (2019) 012012 doi:10.1088/1757-899X/522/1/012012

Improving the Resistance of Cement-Based Composites to Sulphur Dioxide

N Žižková[1,2], L Nevřivová[1], M Lédl[1], and Š Keprdová[1]

[1]Brno University of Technology, Faculty of Civil Engineering, Department of Technology of Building Materials and Components, Veveří 95, 602 00 Brno, Czech Republic

[2]Corresponding author - e-mail: zizkova.n@fce.vutbr.cz

Abstract. The paper describes an investigation into the influence of a crystalline additive on the sulphur dioxide resistance of cement-based composites. Cement mortars made with the assistance of secondary crystallization were exposed to an aggressive environment and tested for ways how the crystalline additive (CA) influences their degradation. The results indicate that the crystalline additive improved the resistance of the cement-based composites to sulphur dioxide attack.

1. Introduction

Cement-based composites, which are materials with an alkaline matrix, are vulnerable to corrosive attack. Corrosive gases of acidic character are a common corrosive medium for all types of concrete structures. Most of these corrosive gases are product of industrial pollution [1]. Industrial combustion of fuel with a high sulphur content in generates oxidized sulphur compounds (SO_x) [2]. SO_2, produced by fuel or gasoline combustion, appears to be one of the most corrosive atmospheric pollutants [3]. The deteriorative effect of sulphur dioxide on concrete is called sulphation [1]. The impact of sulphur dioxide (SO_2) exposure on cement-based composites was described by Matousek and Drochytka [4] and is shown in Figure 1.

The reaction of SO_2, water, sulphurous acid and sulphuric acid with free $Ca(OH)_2$, CaO, dicalcium silicate ($2CaO \cdot SiO_2$) and tricalcium aluminate ($3CaO \cdot Al_2O_3$) forms crystals or precipitates in the concrete matrix with a volume of 5 – 30 times greater than the initial volume of these crystals [2]. The increase in the volume of the crystals increases internal pressure and eventually causes the loss of strength and loss of compactness.

One of the progressive technologies for protecting cement-based composites against aggressive environments is the use of coatings, sprays or additives known as “secondary crystallization” or “crystalline technology” [5]. Laboratory studies [6] have shown that CA as additives contribute to crack healing.

The study [7] confirmed an improvement in the resistance of cement-based mortars containing CA against cyclic freezing as well as gaseous CO_2. An important parameter that influences the effectiveness of the crystalline additive is the curing conditions during the first 28 days of ageing. The research [8] was focused on defining optimum curing conditions. The results showed that the best curing conditions for mortars containing CA are 95 % relative humidity and a temperature of 23 °C. The addition of

Content from this work may be used under the terms of the Creative Commons Attribution 3.0 licence. Any further distribution of this work must maintain attribution to the author(s) and the title of the work, journal citation and DOI.
Published under licence by IOP Publishing Ltd

polypropylene fibres can be harmful to the porosity of hardened mortars, while a higher CA content, i.e. 1.5 %, can bring about a reduction in the porosity of mortars exposed to aggressive environments. The research [9] confirmed the positive influence of a crystalline additive on the porosity of mortars containing polypropylene fibres as well as on the values of compressive strength after exposure to aggressive environments: sulphate solution (concentration of 36 000 mg·L^{-1} of SO_4^{2-}), ammonium chloride solution (concentration of 3 000 mg·L^{-1} of NH_4^+) and gaseous CO_2 at the concentration of φ 65 % and 75 % relative humidity.

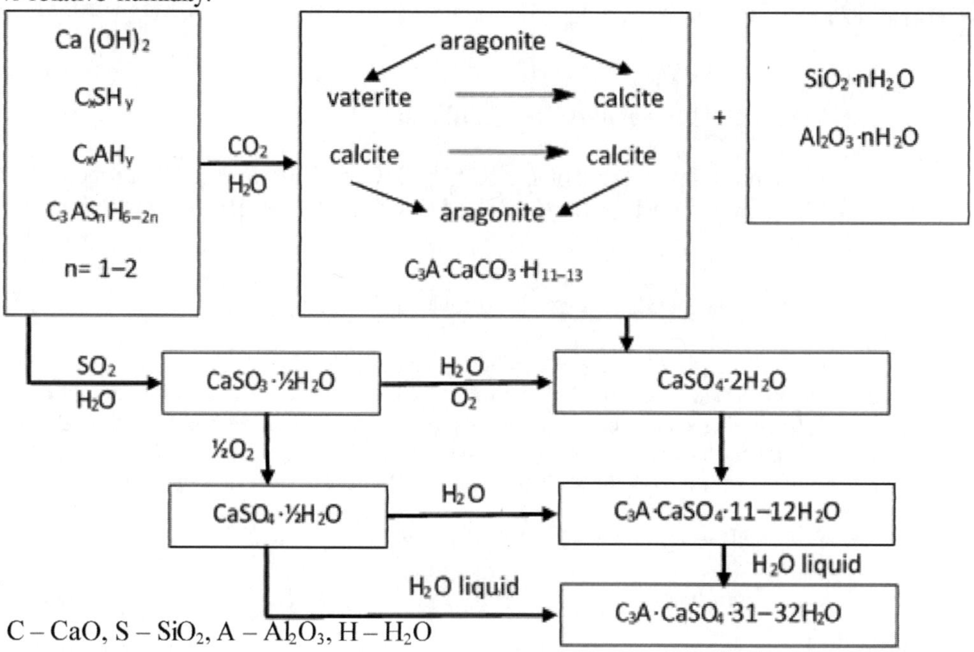

Figure 1. A simplified diagram: final products of concrete sulphation and carbonation [4].

2. Experiment

The goal of this experiment was to investigate the influence of a crystalline additive (CA) on the resistance of cementitious composites to gaseous SO_2. Mortar specimens were made with the dimensions of 40×40×160 mm and were tested for physical-mechanical properties and examined to study their microstructure.

Table 1 shows the composition of the mixtures. Six sets of samples were prepared with Portland cement CEM I 42.5 R as binder, crystalline additive, quartz sand and polypropylene fibres. Crystalline additive was added in an amount of 0.5 and 1.5 % of cement weight, see Table 1.

Table 1. Composition of mixtures.

Mixture ID	OPC [g]	Sand [g]	CA [g]	PP Fibres [g]	Water/Cement ratio
S/I			–	–	0.5
S/II			2.25	–	0.5
S/III	450	1350	6.75	–	0.5
SF/I			–	0.9	0.5
SF/II			2.25	0.9	0.5
SF/III			6.75	0.9	0.5

All the specimens were stored for 28 days at a temperature of 23 °C and relative humidity of 95 %, and subsequently exposed for 90 days to the following 18 cycles: 96 hours at a temperature of 40 °C in gaseous SO_2 at the concentration of $\varphi = 2.2$ % and saturated water vapour in a corrosion chamber and then for 24 hours at a temperature of 23 °C and relative humidity of 50 %.

XRD analysis was performed using a Panalytical EMPYREAN and SEM imaging using a TESCAN MIRA3 XMU.

3. Results and Discussion

The results of compressive and flexural strength tests performed at an age of 118 days (28 + 90) are shown in Figure 2 and 3. The compressive strength of mortars aged at a temperature of 23 °C and relative humidity of 95 % and tested after 18 SO_2 cycles does not show a marked difference, see Figure 2. Flexural strength increases mainly in mortars containing 1.5 % of CA.

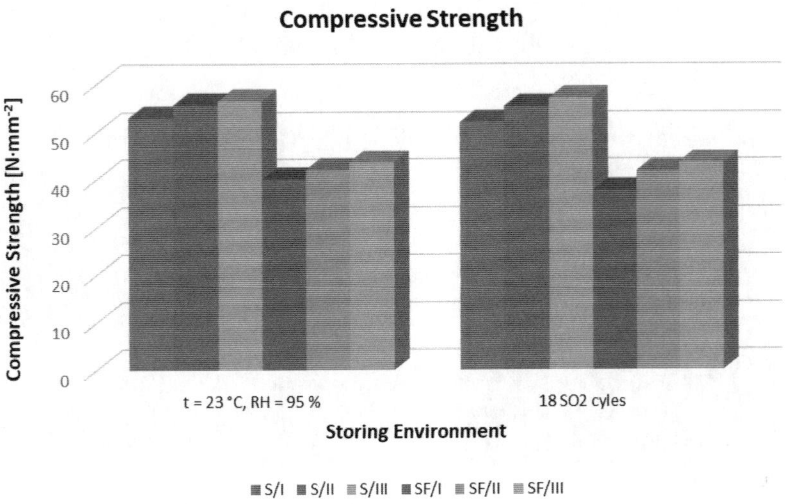

Figure 2. Determination of 118-day compressive strength.

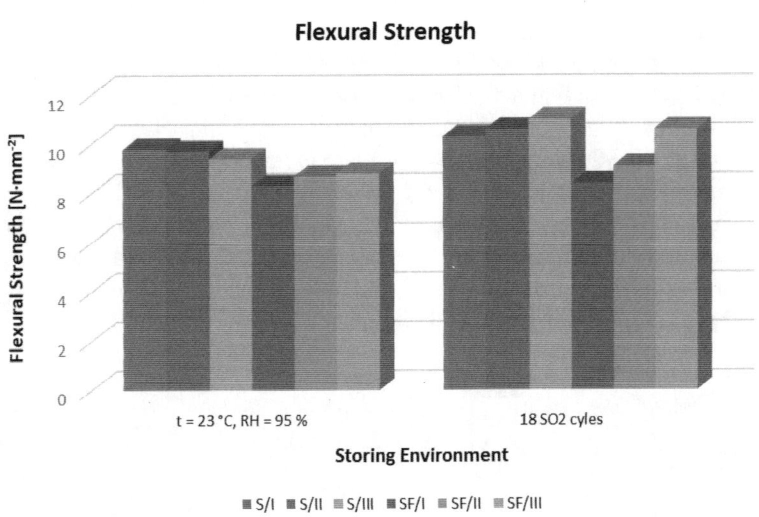

Figure 3. Determination of 118-day flexural strength.

IOP Conf. Series: Materials Science and Engineering **522** (2019) 012012 doi:10.1088/1757-899X/522/1/012012

Once the 18 SO_2 cycles were completed, samples of material located 1 cm beneath the specimen surface were taken and analysed by XRD, see Figure 4 and 5.

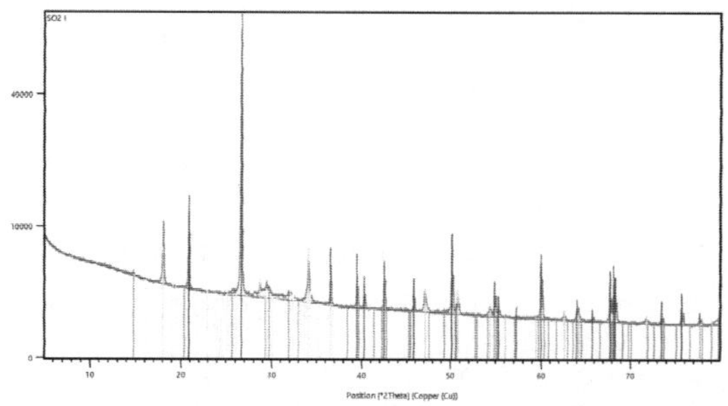

Figure 4. XRD record of mixture S/I (without CA) after 18 SO_2 cycles: SiO_2 (blue), $Ca(OH)_2$ (green), $CaSO_4 \cdot ½H_2O$ (violet), C_3S (yellow).

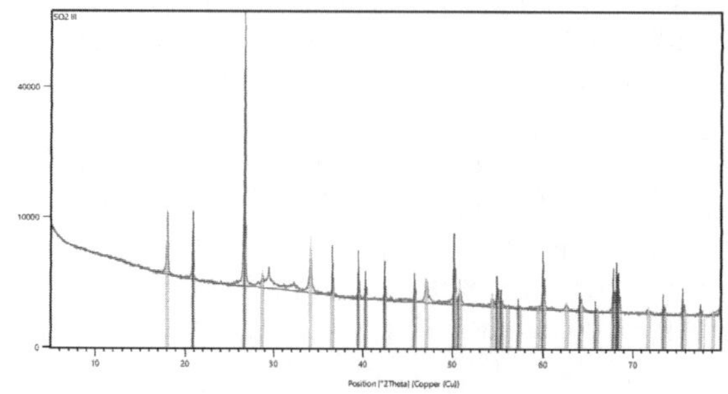

Figure 5. XRD record of mixture S/III (with 1.5 % of CA) after 18 SO_2 cycles: SiO_2 (blue), $Ca(OH)_2$ (green).

XRD analysis identified calcium sulphate hemihydrate ($CaSO_4 \cdot ½H_2O$) in mixture S/I (without CA).

Figure 6. Surface detail of mixture S/I (without CA) after 18 SO_2 cycles.

Figure 7. Surface detail of mixture S/III (with CA) after 18 SO_2 cycles.

Figure 6–7 show a detail of the surface of samples subjected to 18 SO_2 cycles. While sample S/I (without CA) suffered corrosion to the depth of 3 mm, sample S/III (with 1.5 % of CA) corroded down to 0.3 mm. Sample S/I had its surface layer removed and analysed by XRD, which identified calcium sulphate hemihydrate ($CaSO_4 \cdot ½H_2O$) and calcium sulphate ($CaSO_4$).

Figure 8–11 show differences in the microstructure of samples S/I and S/III exposed to 18 SO_2 cycles. Authors [10] demonstrated that SO_2 is rapidly absorbed into the concrete surface, forming sulphate and sulphite species. Our study has found that the surface of samples exposed to 18 SO_2 cycles shows two sulphate formations: calcium sulphate hemihydrate ($CaSO_4 \cdot ½H_2O$) and calcium sulphate ($CaSO_4$). The chemical corrosion mechanism of SO_2 occurring in cementitious composites was described before [11]. This process begins first with the transformation of oxidized sulphur compounds in sulphurous acid (H_2SO_3) and sulphuric acid (H_2SO_4), as follows [11]:

$$SO_2\,(g) + H_2O\,(aq) \rightleftarrows H_2SO_3\,(aq) \quad (1)$$

$$H_2SO_3 + ½O_2 \longrightarrow H_2SO_4 \quad (2)$$

IOP Conf. Series: Materials Science and Engineering **522** (2019) 012012 doi:10.1088/1757-899X/522/1/012012

Figure 8. Picture of mixture S/I (without CA) after 18 SO_2 cycles.

Figure 9. Detail of newly formed (CaSO4·½H2O) crystals.

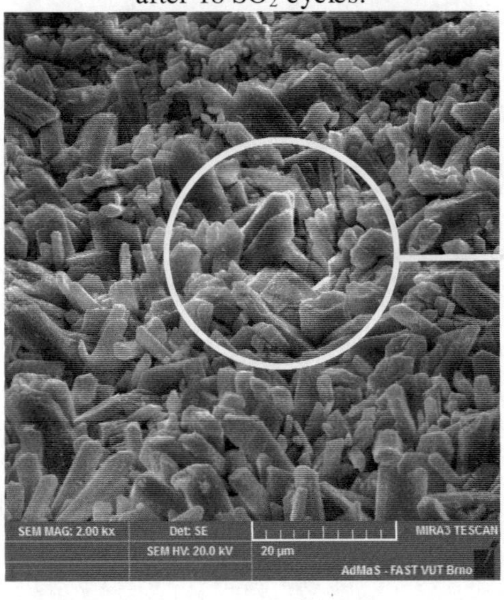

Figure 10. Picture of the surface of mixture S/I (without CA) after 18 SO_2 cycles.

Figure 11. Detail of newly formed ($CaSO_4 \cdot ½H_2O$) crystals on the surface.

The reactions of SO_2, water, H_2SO_3 and H_2SO_4 with calcium hydroxide ($Ca(OH)_2$), calcium oxide (CaO), dicalcium silicate ($2CaO \cdot SiO_2$), and tricalcium aluminate ($3CaO \cdot Al_2O_3$) are forming crystals or precipitates [11]. Newly forming crystals such as corrosion products tend to have a large volume than the original compounds, which causes substantial crystalline pressure inside the cement matrix resulting in damage. One of these corrosion products is calcium sulphate dihydrate ($CaSO_4 \cdot 2H_2O$), which is a calcium slat with a crystallization pressure of 28.2 MPa. The sample without a crystalline additive examined herein contained calcium sulphate hemihydrate ($CaSO_4 \cdot ½H_2O$). Calcium sulphate hemihydrate ($CaSO_4 \cdot ½H_2O$) reacts by the following reaction:

$$2(CaSO_4 \cdot ½H_2O) + 3H_2O \longrightarrow 2(CaSO_4 \cdot 2H_2O) \quad (3)$$

Another possible product is monosulphate ($3CaO \cdot Al_2O_3 \cdot 3CaSO_4 \cdot 12H_2O$) and ettringite ($3CaO \cdot Al_2O_3 \cdot 3CaSO_4 \cdot 32H_2O$). Monosulphate and ettringite, much like calcium sulphate dihydrate ($CaSO_4 \cdot 2H_2O$) are corrosion products with a larger volume and thus present a risk of increased internal pressure within the cement matrix, which causes cracking, deformation and loss of strength.

Calcium sulphate hemihydrate ($CaSO_4 \cdot ½H_2O$) was found on the surface of samples subjected to 18 SO_2 cycles. Samples containing CA sufferred 90 % less severe corrosion than samples without CA. XRD analysis (of material located 1 cm beneath the specimen surface) identified calcium sulphate hemihydrate ($CaSO_4 \cdot ½H_2O$) only in mixture S/I, which contained no CA.

4. Conclusion

The experiments confirmed that the SO_2 resistance of cementitious mortars can be improved by the addition of a crystalline additive. Concerning strength, the flexural strength of mortars containing the crystalline additive increased. No influence on compressive strength was confirmed. However, the study of microstructure showed an improvement in the SO_2 resistance of the mortar containing the crystalline additive.

Acknowledgment

The paper was produced with the financial support of Czech Science Foundation, project No. 16-25472S "Secondary crystallization modified cement composites and their degradation dynamics".

References

[1] Moncmanová A et al. 2007 *Environmental Deterioration of Materials* (Boston: WIT Press)
[2] Mainer F B, Almeida P C F, Nani B, Fernandes L H and Reis M F 2015 *Open Journal of Civil Engineering* **5**, pp 379–389
[3] Yeomans S 2004 *Galvanized Steel Reinforcement in Concrete* (Oxford: Elsevier Science)
[4] Matousek M and Drochytka R 1998 *Atmospheric Corrosion of Concrete* (Prague: IKAS)
[5] Hewlett P C *Lea's Chemistry of Cement and Concrete* (Oxford: Elsevier Ltd. 2010)
[6] Sisomphon K, Copuroglu O and Koenders E A B 2012 *Cement and Concrete Composites* **34**, pp 566–674
[7] Žižková N, Nevřivová L, Lédl M 2017 *Defect and Diffusion Forum* **382**, pp 246–253
[8] Žižková N, Nevřivová L, Lédl M 2017 *Solid State Phenomena* **274**, pp 53–59
[9] Žižková N, Nevřivová L, Lédl M and Keprdová Š 2018 *Material Science and Engineering* **385**
[10] Ramakrishan G, Wu Q, Moon J and Orlov A 2017 *Chemical Engineering Journal* **319**, pp 57-64
[11] Mainer F B et al. 2015 *Open Journal of Civil Engineering* **05**, pp 378–389

IOP Conf. Series: Materials Science and Engineering **522** (2019) 012013 doi:10.1088/1757-899X/522/1/012013

The Influence of Crystallization Additives on the Porosity of Polypropylene Fibre-reinforced Cement Mortar

L Nevřivová[1], **N Žižková**[2]

[1,2] University of Technology, Faculty of Civil Engineering, Department of Technology of Building Materials and Components, Veveří 95, 602 00 Brno, Czech Republic

E-mail: nevrivova.l@fce.vutbr.cz

Abstract. The phenomenon of self-healing cement materials has been described in the professional literature for decades. One self-healing mechanism is based on the presence of crystallization additives in concrete or mortar; when a defect appears in the material, the crystallization additives (CA) are activated, and new crystallization products will fill the pore or crack. The activation of the crystallization additives occurs only under certain conditions. The intensity of crystallization is affected by the nature of the environment within the crack or pores, and in order for the crystallization to occur, the cracks or pores must contain an aqueous solution with sufficient ion content to react with the crystallization additive. This paper discusses the possibility of using crystallization additives to reduce the porosity of cement mortar reinforced with polypropylene (PP) fibres, describes the effect of crystallization additives on the pore structure of cement mortar, and evaluates the influence of the medium on pore size distribution in cement mortar.

1. Introduction

Crystallization additives are added to concrete or mortar to extend the life of a building's structure. If a defect, crack, or tear occurs, crystallization additives present on the surface of the structure mean that if water enters the crack, the minerals react with $Ca(OH)_2$ to form new crystals, gradually filling the defect [1]. The presence of these mineral additives in the mortar ensures secondary crystallization, the products of which fill the defect, and prevents the spread of cracks and the entrance of aggressive substances to the structure, thereby helping to avoid decreases in strength during the life cycle of the structure. For the crystallization additive, it is common to use carbonates of $NaHCO_3$ or Na_2CO_3 [2], talc [3], or sodium silicate [4].

However, use of CA creates a number of problems. The mineral additive is applied directly to the concrete or cement mixture without any protection and soon as it comes into contact with water, the mixture immediately begins to react [5]. As a result, these mineral additives are used prior to the hardening of the cement material, and so their presence in the concrete loses importance, where self-healing is concerned. In order to keep the additive unreacted in hardened concrete, it is necessary to modify it before mixing so as to avoid an immediate reaction with water. Various techniques for the treatment of CA [6, 7, 8] have been described in the literature. The basic condition for secondary crystallization is the constant presence of water; the ions contained in the water exert a considerable influence on the physical and chemical processes, and affect the rate and intensity of the secondary crystallization. This issue is not yet described in the literature, but this fact should be taken into account in further research on self-healing in cement materials.

Content from this work may be used under the terms of the Creative Commons Attribution 3.0 licence. Any further distribution of this work must maintain attribution to the author(s) and the title of the work, journal citation and DOI.
Published under licence by IOP Publishing Ltd

In this case of the present study, these CA were used to reduce the porosity of a PP-reinforced cement mortar in order to eliminate problems with the premature reaction of the crystallization additive. The PP fibres in the mixture reduce stress in the initial phase of the solidification, reduce plastic sedimentation, and prevent microcracks from forming in the first phase of solidification. The fibre-reinforced mortar also exhibits a higher tensile bending strength. However, the addition of fibres causes an increase in the apparent porosity of the material, a decrease in the bulk density, and thus a decrease in compressive strength [9]. The fibres have both a positive and a negative influence on the material.

2. Materials and methods

The experiment described here employed a secondary crystallization additive designed for cement-based mortars. The specimens were divided into two groups; the first batch of specimens was used to observe the effect of a CA, and the second to study the influence of the CA in combination with polypropylene fibres.

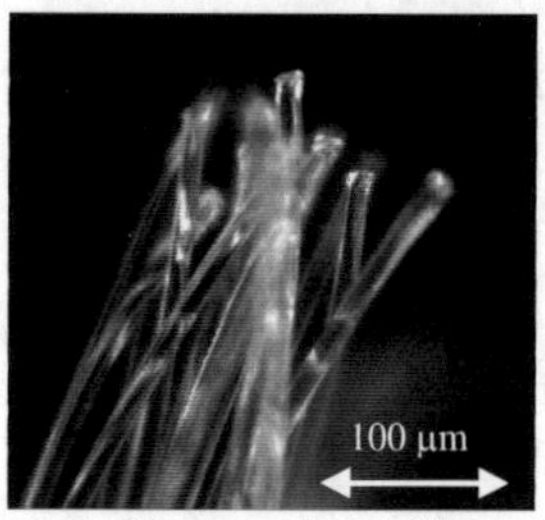

Figure 1. Fibrin polypropylene fibres, with a fibre diameter of 20 micrometres

The cement used as a binder was Portland cement CEM I 42.5 R (Českomoravský cement a.s., Mokrá factory). Sand was used according to EN 196-1, and XYPEX Admix (XYPEX CHEMICAL CORPORATION, Richmond, B.C., Canada) was used as the crystallization additive. The second batch of specimens was made with polypropylene fibres Fibrin 3/15 (KrampeHarex CZ, s.r.o.), Fig 1.

The composition of the mortars tested is shown in Table 1. PP fibres were dosed at 0.0%, 0.5%, and 1.5 % by weight of cement, and the fibre content in the formulation was 0.05%.Please follow these instructions as carefully as possible so all articles within a conference have the same style to the title page. This paragraph follows a section title so it should not be indented.

Table 1. The composition of the mortars

<table>
<tr><th>Mixture ID</th><th>Cement CEM I 42.5 R [g]</th><th>Sand [g]</th><th>CA [g]</th><th>PP Fibres [g]</th><th>Water/cement ratio</th></tr>
<tr><td>REF</td><td rowspan="6">450</td><td rowspan="6">1350</td><td>0.0</td><td rowspan="3">0.0</td><td rowspan="6">0.5</td></tr>
<tr><td>0.5</td><td>2.25</td></tr>
<tr><td>1.5</td><td>6.75</td></tr>
<tr><td>REF*</td><td>0.0</td><td rowspan="3">0.9</td></tr>
<tr><td>0.5*</td><td>2.25</td></tr>
<tr><td>1.5*</td><td>6.75</td></tr>
</table>

The pore structure of the specimens was determined at the following ages, and after being treated by the following curing techniques:

- **H_2O** 28+90days (the specimens were stored at a temperature of 23°C and a relative humidity of 95% for 28 days, and then for another 90 days).
- **SO4** 28+90days in an aqueous solution of Na_2SO_4 (the specimens were stored at a temperature of 23°C and 95% relative humidity for 28 days, after which they were moved into the Na_2SO_4 solution with a concentration of 36,000 mg/l SO_4ions, where they were left for 90 days).
- **NH_4** 28+90 days in an aqueous solution of NH_4Cl (the specimens were stored at a temperature of 23°C and 95% relative humidity for 28 days, after which they were moved into the NH_4Cl solution with the concentration of 3,000 mg/l NH_4ions, where they were left for 90 days).

- **Lab** 28+180 days (the specimens were stored at a temperature of 23°C and 95% relative humidity 28 days, after which they were moved into laboratory conditions with e temperature pf 20°C and 55% relative humidity).

Apparent porosity, average pore diameter, bulk density, and pore size distribution were determined by high-pressure mercury intrusion porosimetry, using a ThermoFinnigan POROTEC of the Pascal 140 – 240 type. Bulk density, apparent density, and apparent porosity were determined using the hydrostatic gravity test (ČSN EN 993-1).

3. Selected results and discussion

The results obtained will be discussed here, together with the results published in [10]; namely, the properties of the cement mortar prepared according to the same formulation and stored for 28+180 days in a laboratory environment. These samples were labelled Lab or Lab*, depending on whether they contain PP fibres.

Firstly, the effect of PP fibres on the pore structure of the cement mortar is assessed. Although the fibres are only dosed at 0.05 wt. % on the cement, and their effect on the open pore content of the material is clearly visible. The apparent porosity of the mortar increased by up to 43%, as shown in Table 2. The increase in porosity is related to the deterioration of the cement mortar's workability, and is also due to the fact that air is being drawn into the mixture with the fibres, and that the air cannot escape from the mixture.

Table 2. Results of the mercury intrusion porosimetry test

	H_2O						SO_4					
RESULTS	**REF**	**0.5**	**1.5**	**REF***	**0.5***	**1.5***	**REF**	**0.5**	**1.5**	**REF***	**0.5***	**1.5***
Average pore diameter (μm)	0.081	0.083	0.087	0.090	0.100	0.085	0.103	0.078	0.084	0.073	0.092	0.088
Apparent porosity (%)	11.9	11.1	10.8	15.9	14.7	13.8	10.5	10.6	9.6	14.7	12.9	12.3
Porosity increase (%)				**34.1**	**32.9**	**28.3**				**40.1**	**22.2**	**28.9**
Bulk density (g/cm³)	2.176	2.213	2.176	2.076	2.016	2.105	2.207	2.232	0.450	2.081	2.141	2.148
Apparent density (g/cm³)	2.442	2.488	2.499	2.469	2.364	2.486	2.466	2.495	0.460	2.439	2.458	2.450
	NH_4						Lab					
RESULTS	**REF**	**0.5**	**1.5**	**REF***	**0.5***	**1.5***	**REF**	**0.5**	**1.5**	**REF***	**0.5***	**1.5***
Average pore diameter (μm)	0.082	0.085	0.084	0.084	0.102	0.092	0.103	0.102	0.099	0.102	0.103	0.075
Apparent porosity (%)	10.3	10.5	9.2	14.7	13.8	12.3	13.8	13.5	13.2	16.2	16.3	16.3
Porosity increase (%)				**42.8**	**39.5**	**33.3**				**17.4**	**21.1**	**24.0**
Bulk density (g/cm³)	2.187	2.223	2.228	2.108	2.131	2.177	2.151	2.172	2.091	2.092	2.112	2.080
Apparent density (g/cm³)	2.401	2.485	2.455	2.471	2.499	2.483	2.511	2.510	2.407	2.497	2.524	2.486

The difference in porosity between the PP-reinforced and non-reinforced mortar is shown in Fig. 2 below, while it is also possible to assess the influence of the environment in which the samples were stored. The apparent porosity of the samples is highest when stored in a lab environment (Lab), even though those samples are 90 days older than the others. The porosity varies between 13-14% for the fibre-free specimens and 16% for the fibre specimens. When stored at a relative humidity of 95% (H2O), the porosity is lower, about 11% in the samples without PP fibres, and 14-16% in the fibre samples. The lowest porosity was found in the samples stored in an aqueous solution (SO4, NH4). The apparent porosity is 9-10.5% for non-reinforced mortar and a maximum of 14.7% for reinforced mortar.

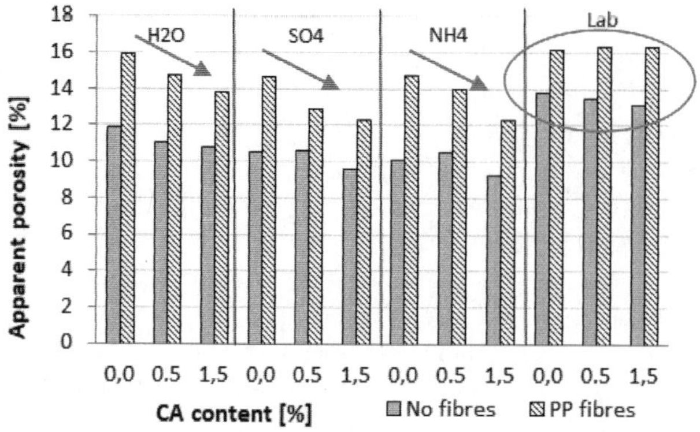

Figure 2. The effect of the crystallization additive on the apparent porosity of a mortar being hardened in different environments

The addition of fibre causes an increase in pore content of 1.0-0.05 μm in all environments, as shown in Fig. 3. These pores are then filled with secondary crystallization products. In the laboratory environment, this secondary crystallization almost does not occur because there is insufficient moisture in the pores, which is necessary for the reaction of the crystallization additive, as shown in Fig. 2. The dose of the crystallization additive did not influence porosity in the laboratory environment. A different result was observed for the mortar stored at 95% relative humidity. The addition of CA reduced the apparent porosity in the samples with PP fibres to 13%, while in the case of fibre-free samples, the porosity is even lower. The same trend was also seen for the mortars stored in aqueous solutions (SO4 and NH4).

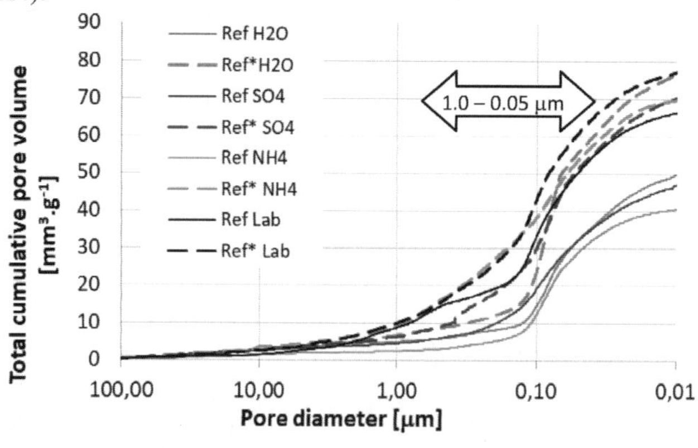

Figure 3. The effect of crystallization additive on pore size distribution

One example of this is the decrease in apparent porosity in the reinforced and non-reinforced mortar deposited in NH_4Cl solution. The apparent porosity of PP-free mortar was reduced from 10.3% to 9.2%, and with the fibre mortar from 14.7% to 12.3%. The reaction products gradually fill up the pores that have been created in the mortar through the addition of PP fibres, i.e. pores sized 1-0.05 μm. The higher the CA dose of the additive, the higher the crystallization products and the lower the porosity, as can be seen in Fig. 4.

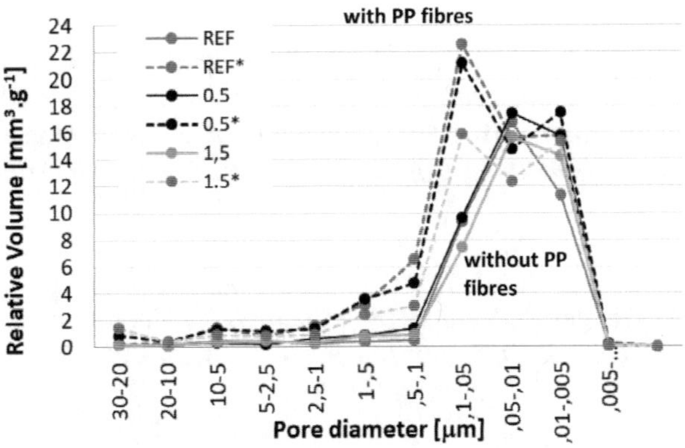

Figure 4. The pore size distribution of cement mortar being hardened in an aqueous solution of NH_4Cl

4. Summary

The present study examined the effect of the crystallization additive XYPEX on the pore structure of cement mortar. This is a mineral additive that is primarily used for the self-healing component of cement materials. In our case, its influence on the apparent porosity of mortar reinforced with PP fibre has been verified.

The cement mortar was prepared according to a standard procedure; the samples were stored for 28 days at a relative humidity of 95% and subsequently transferred to different environments, where they were hardened for 90 and 180 days, respectively.

The pores formed in the mortar after the addition of PP fibres can be reduced by adding the XYPEX crystallization additive. The effectiveness of the additive depends on its dosing, as well as on the setting and hardening conditions of the cement mortar. This study has also verified that the use of an additive is not suitable for laboratory samples that do not allow the crystallization reaction to start. Conversely, a high relative humidity or aqueous solution environment is suitable for use with XYPEX; the CA dose of 1.5% by weight of cement reduced the porosity of the PP fibre mortar by up to 27%. The porosity of 12.3% found in this case approximates the porosity of the unreinforced mortar.

Acknowledgment

The paper was produced with the financial support of Czech Science Foundation, project No. 16-25472S “Secondary crystallization modified cement composites and their degradation dynamics”.

References

[1] K Sisomphona, O Copuroglu and E.A.B. Koenders 2012 Self-healing of surface cracks in mortars with expansive additive and crystalline additive, *Cement and Concrete Composites* Vol. 34 (4) p. 566-574.

[2] T Kishi, T H Ahn, A Hosoda, S Suzuki and H Takaoka 2007 Self-healing behaviour by cementitious recrystallization of cracked concrete incorporating expansive agent. In Proceedings of *1st International Conference on Self Healing Materials.*

[3] M Roig-Floresa, S Moscatob, P Serna and L Ferrara 2015 Self-healing capability of concrete with crystalline admixtures in different environments. In *Construction and Building Materials* Vol.86 (0) p. 1-11.

[4] G Yildirima, M Sahmaran, M Balcikanli and E Ozbay 2015 Influence of cracking and healing on the gas permeability of cementitious composites. In *Construction and Building Materials* Vol.85 (0), p. 217-226.

[5] R Sahamitmongkol, T Kishi 2011 Tensile behavior of restrained expansive mortar and concrete, In *Cement and Concrete Composites* Vol 33 (1), p. 131-141.

[6] V V Hung, T Kishi and T H Ahn 2013 Development of self-healing granules having semi-capsulation effect by using cement compound, chemical/mineral admixtures & its watertight

performance through crack, Proceedings of the 4th International *Conference on Self-Healing Materials* ICSHM 2013.

[7] H Dong, H Huang and G Ye 2013 Inorganic powder encapsulated in brittle polymer particles for self-healing cement based materials, 4th *International conference on Self-healing Materials.*

[8] Y-S Lee and J-S Ryou 2014 Self healing behavior for crack closing of expansive agent via granulation/film coating method. In *Construction and Building Materials* Vol. 71 (0), p. 188-193.

[9] N Žižková, L Nevřivová and M Lédl 2018 Durability of Cement Based Mortars Containing Crystalline Additives, *Defect and Diffusion Forum*, Vol. 382, pp. 246-253.

[10] L Nevřivová, N Žižková and M. Lédl 2018 Influence of a Crystalline Additive on the Porosity of Cement Mortars, *Defect and Diffusion Forum*, Vol. 382, pp. 254-260.

Numerical Analysis of Solid Fuel Regression Rate Distribution Characteristics in Solid-fuel Ramjet with Different Fuel Types

Omer Musa[1, 2, a], Chen Xiong[1,b], Liao Wenhe[1,c] and Li Yingkun[1,d]

[1]School of Mechanical Engineering, Nanjing University of Science and Technology, Nanjing, China
[2]Mechanical Engineering Department, Omdurman Islamic University, Sudan

[a]omer.musa1@hotmail.com, [b]chenxiongnjust@njust.edu.cn, [c]52017083@njust.edu.cn, [d]liyingkun@njust.edu.cn

Abstract. This paper proposes and examines numerically a new design of solid-fuel ramjet motor. Multi-physics coupling code is developed using FORTRAN and parallel computing to solve the problems of multi-physics coupling of fluid mechanics, solid pyrolysis, heat transfer, thermodynamics, and chemical kinetics. Simulations are carried out for the proposed design then the results are compared with the classic design of the solid-fuel ramjet. It is found that the proposed design has improved the regression rate significantly; besides, the amount of released solid fuel is increased for the same size.

1. Introduction

The combination of range and speed is the most needed feature in the warfighters. Traditional rocket motor requires carrying an oxidizer which makes the rocket heavier and then decreases its flight range. For the same size of rocket motor, solid-fuel ramjet (SFRJ) could flight for longer range with much higher speed since it uses its forward motion to draw in oxygen from the surrounding atmosphere. Therefore, the larger fuel store allows SFRJ to sustain high speed during flight. Besides, SFRJ is a simple air-breathing engine, in design, that contains no moving parts. Thus, the internal design of the SFRJ determines its performance.

However, low fuel regression rate is the main disadvantage of the SFRJ, usually less than one mm/sec [1], which is the key parameter in the design and operation of SFRJs that is primarily depend on the heat transfer from the flame to the fuel surface. The classical design of SFRJ basically consists of air intake system, combustion chamber, and nozzle in which the combustor contains flame-holder and cylindrical solid fuel grain (see Fig. 1 - upper). However, in order to overcome the disadvantage of SFRJ and enhances its combustion characteristics and performance we proposed and studied a new SFRJ design which contains additional solid fuel (rod) at the center of the tubular grain (see Fig. 1 - lower). The idea is to increase the solid fuel surface area contacting with incoming air to enhance the heat transfer to the solid fuel and then the regression rate. Besides, maintaining the amount of solid fuel released during flight which makes the thrust stable. In this paper, the proposed SFRJ design is investigated numerically using an in-house CFD code. The code is written by FORTRAN with parallel computing to solve the problems of multi-physics coupling of fluid mechanics, solid pyrolysis, heat transfer, thermodynamics, and chemical kinetics.

In open literature, few studies are available regarding different solid-fuel ramjet combustor's designs. However, complex configurations have been investigated extensively in hybrid rocket motors. Lee et al.

Content from this work may be used under the terms of the Creative Commons Attribution 3.0 licence. Any further distribution of this work must maintain attribution to the author(s) and the title of the work, journal citation and DOI.
Published under licence by IOP Publishing Ltd

[2] studied the regression rate characteristics of helical grain. A higher regression rate is achieved when the pitch number is small, and the combination of helical grain with a large pitch number and a moderate inlet swirl may produce the maximum regression rate. Li et al. [3] investigated numerically the regression rate distributions for tube, star, and wagon wheel grains. The regression rates of star and wagon wheel grains were generally higher than that of tube grain under the same oxidizer mass flux.

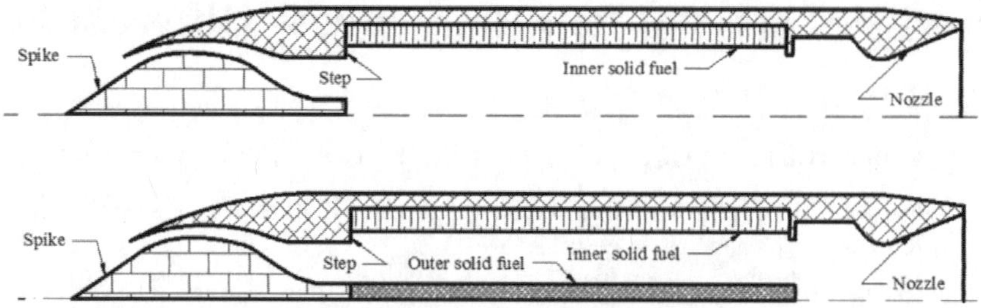

Figure 1. The classic (upper) and proposed (lower) design of solid-fuel ramjet.

2. Mathematical Modeling

2.1. Governing Equations.

The flow field considered is essentially three dimensional (3-D) for which all velocity and Reynolds stress components are included with regard to the simplifying condition of symmetry, whereas the flow is still angular symmetry *(∂/∂θ=0)*. Therefore, the 3-D Reynolds-average Navier-Stokes equations with angular symmetry condition in integral form, which are representing the conservation of mass, momentum, energy, and species equations, can be written as:

$$\frac{\partial}{\partial t}\iiint_{\Omega} U d\Omega + \iint_{s}\left(\boldsymbol{F}\cdot n_x + \boldsymbol{G}\cdot n_y\right)ds - \iint_{s}\left(\boldsymbol{F}_v\cdot n_x + \boldsymbol{G}_v\cdot n_y\right)ds = \iiint_{\Omega}(\boldsymbol{H}+\boldsymbol{H}_v).\varepsilon d\Omega + \boldsymbol{S} \quad (1)$$

where $n = n_x i + n_y j$ is the unit normal vector in outward direction of the boundary surface s ; t is the time, $\varepsilon = 0$ for two-dimensional plane flow, $\varepsilon = 1$ and for axisymmetric flow. The conservative vector $\boldsymbol{U}$, convective flux vectors $\boldsymbol{F}$, $\boldsymbol{G}$, viscous flux vectors $\boldsymbol{F_V}$, $\boldsymbol{G_V}$, axisymmetric source terms $\boldsymbol{H}$, $\boldsymbol{H_V}$, and chemical reactions source term $\boldsymbol{S}$ can be found elsewhere [4].

Arrhenius formula (Eq. 2) is adopted to determine reaction rate constants.

$$\dot{r}_k = A_k T_w^{n_k} \exp\left[-(E_a)_k / R_u T_w\right] \quad (2)$$

where A_k is the pre-exponential factor, n_k is the temperature exponent, E_a is the activation energy, and T_w is the solid-fluid interface wall temperature. In order to model the solid fuel pyrolysis in the solid-fluid interface the zeroth order ($n_k = 0$) Arrhenius equation is used. The mass rate of fuel freed by pyrolysis is determined using:

$$\dot{m} = \rho_{sol}\dot{r} = \rho_{sol} A_{sol} \exp\left[-(E_a)_{sol} / R_u T_w\right] \quad (3)$$

The subscript “sol” indicates the solid fuel. The solid fuel has been used in this work is the high-density Polyethylene (HDPE) grain. The gas-phase chemical reactions kinetic of HDPE [5] can be written as in Table 1.

Table 1. Chemistry Model of Ethylene.

Chemical reaction	$A/(\mathrm{cm^3/mol\cdot s})$	**n**	*Ea(J/mol)*
$C_2H_4 + O_2 \rightarrow 2CO + 2H_2$	2.10E14	0	149779.2
$2CO + O_2 \rightarrow 2CO_2$	3.48E11	2	84261.5
$2H_2 + O_2 \rightarrow 2H_2O$	3.00E20	-1	0.0

where the number of reactions are three (*I=3*) with six species (*N=6*).

2.2. Numerical Solution.

The prescribed governing equations have been solved simultaneously by means of cell-centered, finite-volume method with structured grids, multi-block, and density-based approach. At each cell of the SFRJ flow field meshes (104982 cells) the governing equations are discretized and numerically solved in which AUSMPW+ [6], third-order upwind-based MUSCL [7], and Van Albada limiter function [8] are employed to evaluate inviscid fluxes with the second-order central differences for viscous fluxes. The turbulence is treated by Menter's shear-stress transport (SST) turbulence model [9]. Finite rate model is employed to determine the chemical source terms. In this model the flame is treated as laminar flame with neglecting the effect of turbulent fluctuating (turbulence-chemistry interaction), at which Arrhenius expression is used to determine the mean reaction rate. Since we considered unsteady simulations, the lower-upper symmetric Gauss-Seidel (LU-SGS) implicit dual time-stepping algorithm is selected for fluid domain and backward Euler method for solid domain [10].

In this work, two cases are simulated with different combustor designs, the proposed design and the classic one. The total temperature is set to be 540K and the mass flow rate is fixed at 0.6 kg/s. Other boundary conditions are treated as follows; no-slip adiabatic boundary condition at walls, axisymmetric boundary condition at axis, all axial gradient quantities are set to zero at exit, and at solid-fluid interface the solid and fluid domains are coupled by direct coupling approach in which the wall temperature is determined by energy balance equation.

3. Model Verification and Validation

An extensive verification and validation study has been performed on the base code in our previous work [11], including uncertainty and numerical errors. In addition, the base code has been also validated using swirl benchmark cases in the previous work [12]. Furthermore, the extended code with combustion model needs to be validated on combustion cases, to do so; additional two cases have been studied as well in the previous works [4, 13, 14].

4. Results and Discussion

This section presents the numerical results obtained for the cases prescribed in previous sections. The ignition process is achieved by using hot gas flowing together with the incoming air into the combustor. The ignition gas flows with 0.4 kg/s, total temperature of 2500 K and contains N2, H O2, and CO2. The ignition gas is produced through combustion of double-based propellant and only the gas components are considered in the model. After 0.35 second the ignition gas has stopped and only the incoming air flow is continued.

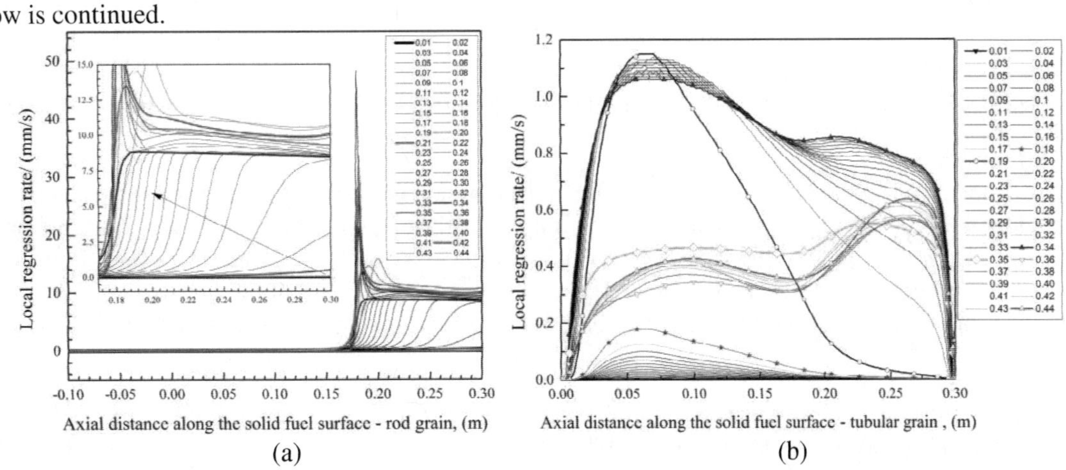

Figure 2. Axial distributions of local regression rate for solid fuel surface at different time intervals.

Normally, ignition delay time includes; heating (gasification) time delay, mixing time delay, and chemical processes time delay; but mixing time is very small and can be ignored. Fig. 2 shows the ignition and combustion processes during the whole simulation for the proposed design. Fig. 2a presents

the local regression rates for outer surface of the rod solid fuel. As can be seen, the pyrolysis starts for the end of the combustor (0.3m) due to the effect of incoming air jet. As time progresses the pyrolysis moved towards the inlet till the distance 0.17 m. For sustained combustion, the regression rate is significantly much higher than tubular grain surface regression. Fig. 2b shows the local regression rates of inner surface of the tubular solid fuel.
On comparing the regression rates of the classical and the new designs, the proposed designs has higher regression rate with a new trend. In which, the peak is shifted into the end of the combustor due to the high temperature offered by the combustion of the outer grain (see Fig. 3). This can be seen clearly in Fig. 4, where the temperature contour represents the flams (high temperature) produced near to each surface.

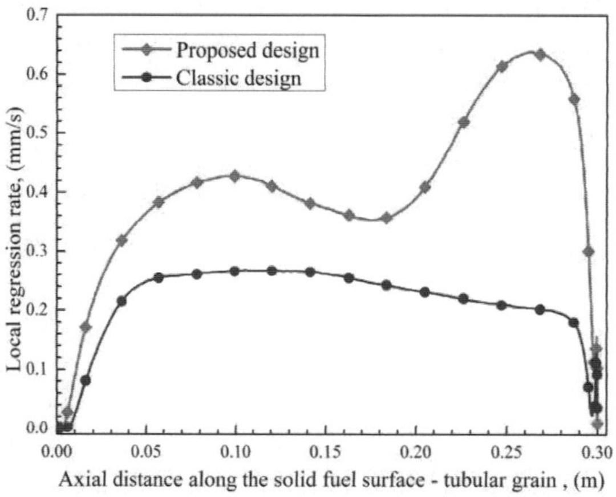

Figure 3. Regression rates along axial direction for the proposed and the classic design.

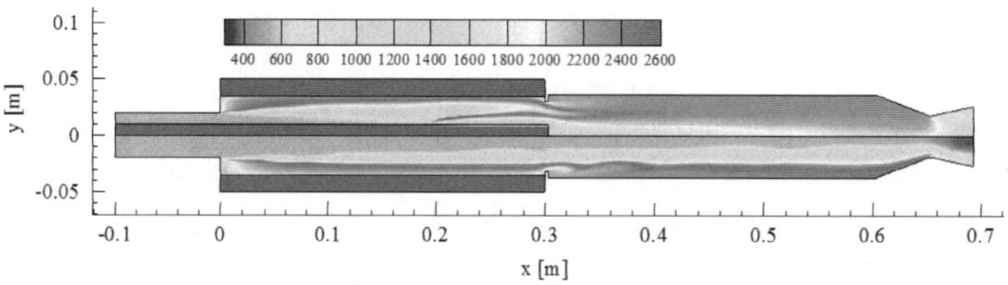

Figure 4. Temperature contours of the proposed (upper half) and the classic design (lower half) [K].

5. Summary

A new design of solid-fuel ramjet is proposed and examined numerically in this paper. CFD code is developed using FORTRAN and parallel computing. The developed code is structured, finite-volume, density-based, cell-centered flow solver that simulates the axisymmetric, unsteady, turbulent, reacting flows. It has been shown that the proposed design has improved the local regression rate and increased the amount of solid fuel inside the motor for the same size compared to the classical solid-fuel ramjet.

References

[1] D. Duesterhaus, A. Hogl, Measurements in a Solid Fuel Ramjet Combustion with Swirl, AIAA Paper (88–3045). doi:10.2514/6.1988-3045.
[2] Lee, C., Na, Y., Lee, J. W., and Byun, Y. H. (2007). "Effect of induced swirl flow on regression rate of hybrid rocket fuel by helical grain configuration. " Aerosp. Sci. Technol., 11(1), 68 –76.
[3] Li, X., Tian, H., and Cai, G. (2013). "Numerical analysis of fuel regression rate distribution characteristics in hybrid rocket motors with different fuel types. " Sci. China Technol. Sci., 56(7), 1807 –1817

[4] O. Musa, C. Xiong, Z. Changsheng, W. Li. 2017. Effect of inlet conditions on swirling turbulent reacting flows in a solid fuel ramjet engine. Applied Thermal Engineering. 113:186-207.

[5] Mawid, M. A., and B. Sekar. 1997. Kinetic modeling of ethylene oxidation in high speed reacting flows. AIAA Paper 3269.

[6] Kim, Kyu Hong, Chongam Kim, and Oh-Hyun Rho. 2001. Methods for the accurate computations of hypersonic flows: I. AUSMPW+ scheme. Journal of Computational Physics 174.1. 38-80.

[7] Van Leer, Bram. 1979. Towards the ultimate conservative difference scheme. V. A second-order sequel to Godunov's method. Journal of computational Physics 32.1. 101-136.

[8] Van Albada, G. D., Bram Van Leer, and W. W. Roberts Jr. 1997. A comparative study of computational methods in cosmic gas dynamics: Upwind and High-Resolution Schemes. Springer Berlin Heidelberg, 95-103.

[9] Menter, Florian R. 1994. Two-equation eddy-viscosity turbulence models for engineering applications. AIAA journal 32.8. 1598-1605.

[10] Zhang, L. P., and Z. J. Wang. 2004. A block LU-SGS implicit dual time-stepping algorithm for hybrid dynamic meshes. Computers & fluids 33.7. 891-916.

[11] O. Musa, Z. Changshen, C. Xiong, L. Yingkun. 2016. Verification Study of a CFD-RANS Code for Turbulent Flow at High Reynolds Numbers. International Journal of Modeling and Optimization. 6. 1-10.

[12] O. Musa, C. Xiong, Z. Changsheng, G. Lunkun. 2016. Assessment of the modified rotation/curvature correction SST turbulence model for simulating swirling reacting unsteady flows in a solid-fuel ramjet engine. Acta Astronautica,. 129. 241–252.

[13] O. Musa, Z. Changsheng, C. Xiong, G. Lunkun. 2016. Prediction of swirling cold flow in a solid-fuel ramjet engine with a modified rotation/curvature correction SST turbulence model. Applied Thermal Engineering. 105. 737-754.

[14] O. Musa, C. Xiong, Z. Chang-sheng, L. Ying-kun, L. Wen-He. Investigations on the influence of swirl intensity on solid-fuel ramjet engine. Computers & Fluids, Volume 167, 2018, Pages 82-99.